Foundations of Engineering Mechanics

Series editors

More information about this series at http://www.springer.com/series/3582

Vladimir Fridman

Theory of Elastic Oscillations

Equations and Methods

Springer

Vladimir Fridman
Los Angeles, CA
USA

Translated by Eugene Sviyazheninov

ISSN 1612-1384 ISSN 1860-6237 (electronic)
Foundations of Engineering Mechanics
ISBN 978-981-10-4785-5 ISBN 978-981-10-4786-2 (eBook)
DOI 10.1007/978-981-10-4786-2

Library of Congress Control Number: 2017940825

Printed on acid-free paper

This Springer imprint is published by Springer Nature
The registered company is Springer Nature Singapore Pte Ltd.
The registered company address is: 152 Beach Road, #21-01/04 Gateway East, Singapore 189721, Singapore

Preface

Since the second half of the last century there has been significant development in the methods used to solve applied problems of mechanics. There are a number of reasons for this. First, as a result of the increasing needs of technology the problems to be researched have become ever more complicated in themselves. Second, there is unprecedented quantitative growth, as well as improvement of the quality of computer equipment and the expansion of its capabilities. Finally, there is the development of computational methods associated with the achievements of mathematical physics and functional analysis.

Techniques based on finite-dimensional approximation of differential and integral equations, such as the methods of finite differences, finite elements, as well as of influence matrices, have become widespread. Their advantage lies in universality and coverage of a wide class of problems. The disadvantages include the need to deal with large numerical arrays. Moreover, by splitting the calculated object into smaller and smaller elements to improve the accuracy of computation, singular matrices can be encountered [10].

The alternative approach, to which this book is devoted, does not differ in universality and requires the development of algorithms to solve each problem or a certain class of problems separately. This includes the so-called variational, projection and iterative methods which, when applied to nonlinear problems, use the procedure of linearization of original nonlinear equations. The composition of efficient algorithms to solve specific individual complicated problems demands a certain skill in the use of mathematical physics and functional analysis. However, significant results can be achieved in this way. According to the variational approach, the equations of the problem correspond to the functionality, extremality or stationarity conditions that are equivalent to the equations themselves. To construct an approximate solution to the problem, functional variation needs to be removed. To do this an approximate solution is represented as a finite series expansion in so-called coordinate functions, and then so-called projection conditions follow from functional stationarity conditions. Projection conditions, approximately replacing the original equations, essentially represent the removal of scalar products from the residual of equations and from the sequence of coordinate functions. The core question consists in choosing a system of coordinate functions. These functions should be linearly independent and possess some properties of the desired solution; for example, to satisfy all or some of the conditions of differentiability and boundary conditions. For this purpose, one can use the spectral properties of the close objective; for example, when considering an elastic body with variable inertial and elastic properties it is convenient to take the forms of free oscillations (natural forms of vibrations) of a homogeneous body as the coordinate functions. For ordinary linear differential equations the procedure based on the use of variational conditions leads to a finite system of algebraic equations. If the selected coordinate functions are not only linearly independent, but also orthonormal in a certain sense, as when using the spectral properties of the nearby problem, then the resulting system of algebraic equations will be well conditioned.

Approximate solution of an extremal problem by reducing it to a system of algebraic equations was first proposed by W. Ritz [9]. B.G. Galerkin [2], [3] pointed to the possibility of composing projection conditions, bypassing their variational justification, which greatly expanded the class of problems that can be solved using projection methods. L.V. Kantorovich extended the application of projection methods to partial differential equations [6]. In nonlinear problems, Newton proposed the iterative method of solution, based on linearization of nonlinear operations by differentiating them by retaining only the first derivative [4]. Cauchy extended the iterative method of Newton to multidimensional extremal problems, using the concept of functional gradient with determination of the step length just from the linearized extremum condition of the functional itself [1]. Thus, Cauchy laid the foundations of iterative gradient methods to solve linear and nonlinear problems. L.V. Kantorovich in his fundamental paper [4] gave the generalized interpretation of approximate methods to solve linear and nonlinear operator equations, based on the notions of functional analysis, thereby significantly extending the class of problems that can be solved using these methods. The most complete statement of approximate methods for solving operator equations is given in Russian in the book [7]. The monograph [8] is devoted to detailed description and justification of variational methods.

This book is written on the basis of the course of lectures delivered over many years at the Physics and Mechanics Faculty of Leningrad Polytechnic Institute. The part of the book dedicated to applied problems was primarily developed at the Institute for Problems in Mechanical Engineering of the Russian Academy of Sciences.

Special thanks are given to Dr. Eugene Sviyazheninov for translating this book from Russian into English.

St. Petersburg, Los Angeles, USA
2016

Vladimir Fridman

References

1. Cauchy, A. (1847). Methode generale pour la resolution des systemes d'equations simultanees. *Comptes rendus de l'Académie des sciences*. Tome 25 (pp. 536–538).
2. Fletcher, C. A. J. (1984). *Computational Galerkin methods* (p. 310). Berlin: Springer.
3. Galerkin, B. G. (1915). *Rods and plates* (Vol. 1, No. 19, pp. 897–908 (in Russian)). Series in some issues of the elastic equilibrium of rods and plates, Herald of engineers.
4. Kantorovich, L. V. (1948). *Functional analysis and applied mathematics* (Vol. 3, No. 6, pp. 89–185 (in Russian)). Successes of Mathematical Sciences, *M. L. Publishing house Ogiz.*
5. Kantorovich, L. V., & Akilov, G. V. (1982). *Functional analysis* (p. 604). Pergamon Press.
6. Kantorovich, L. V., & Krylov, V. I. (1964). *Approximate methods of higher analysis* (p. 681). New York: Interscience Publication.
7. Krasnosel'skii, M. A., Vainikko, G. M., Zabreiko, P. P., et al. (1972). *Approximate solution of operator equations* (p. 496). Springer.
8. Mikhlin, S. G. (1964). *Variational methods in mathematical physics*. Oxford: Pergamon Press.
9. Ritz, W. (1909). *Uber eine neue Methode zur Losung gewisser Variationsprobleme der mathematischen Physik Journal reine und angew, Mathematical (Grelle)* (Bd. 135, No. 1, pp. 1–61).
10. Zienkiewicz, O. C. (1977). *The finite element method* (3rd ed., p. 787). London: McGraw-Hill.

Contents

Summary

Techniques based on finite-dimensional approximation of differential and integral equations, such as the methods of finite differences, finite elements, as well as of influence matrices, have become widespread. Their advantage lies in universality and coverage of a wide class of problems. The disadvantages include the need to deal with large numerical arrays. Moreover, by splitting the calculated object into the smallest elements to improve the accuracy of computation, singular matrices can be encountered.

The book is devoted to the alternative approach to solving the problems of both linear and nonlinear oscillations of elastic distributed parameter systems. This includes the so-called variational, projection and iterative gradient methods which, when applied to nonlinear problems, use the procedure of linearization of the original nonlinear equations. These methods do not differ in universality and require development of a means of solving each problem or a certain class of problems separately. Thus, in many cases more efficient algorithms for the solution of important applied problems can be obtained. To record these algorithms in a unified form, in the first part of the book and its appendix much attention is paid to compilation of general operator equations, which include as particular cases the equations of vibrations of rods, plates, shells and three-dimensional bodies. These are considered mainly periodic or close to periodic oscillations, which correspond to stationary or nearly stationary regimes of machinery operation. In the second part of the book examples of solutions of some applications have been provided.

Summary

Part I
Equations and Methods

Chapter 1
Oscillation Equations of a Rod with Rectilinear Axis

Let us consider the longitudinal, torsional and transverse vibrations of a rod with a rectilinear axis of homogeneous material over its cross-section.

We introduce a rectangular coordinate system xyz, directing the axis x along the axis of the rod, $0 \le x \le l$, where l is the length of the rod.

Assume that each rod section is symmetric relative to the axes y and z. Let us denote by $S(x)$ the cross-sectional area of the rod, $\rho_v(x)$ the specific mass (mass per unit volume), $E(x)$ the elastic modulus in tension of the material, $G(x)$ the shear modulus, $J_p(x)$ the equatorial moment of the rod cross-section, $J(x)$ the cross-sectional moment of inertia about the axis z. We also introduce $\rho = \rho_v S$ as the mass intensity, distributed along the axis of the rod (linear mass, or mass per unit length), $\rho = \rho(x)$.

1.1 Differential Equations of Longitudinal Vibrations of a Rod

We consider that the cross-sections of a rod with a rectilinear axis subject to longitudinal vibrations remain plane and perpendicular to the axis x.

Before turning to compilation of the equations describing longitudinal vibrations of a rod, let us dwell on a more simple static problem.

Suppose that a longitudinal distributed load of intensity $h = h(x)$ acts on the rod, its positive direction coinciding with the positive direction of the axis x. Under the action of this load at any rod cross-section there arise only normal stresses, which are distributed uniformly over the cross-section. Let us denote these stresses as $\sigma = \sigma(x)$. We have

$$\sigma = \frac{Q}{S}, \tag{1.1}$$

V. Fridman, *Theory of Elastic Oscillations*, Foundations of Engineering Mechanics, DOI 10.1007/978-981-10-4786-2_1

where $Q = Q(x)$ is the internal force in the cross-section of the rod with coordinate x. It is directed along the axis x and is positive if it creates positive (i.e. tensile) stresses.

The equilibrium condition of the rod element of length dx is written as

$$(Q + \mathrm{d}Q) - Q + h\mathrm{d}x = 0$$

or

$$-\frac{\mathrm{d}Q}{\mathrm{d}x} = h. \tag{1.2}$$

Displacement of all the points of the cross-section is identical and parallel to the axis x. This displacement $u = u(x)$ is considered to be positive if it occurs in the positive direction of the axis x. The relative extension $\varepsilon = \varepsilon(x)$—the longitudinal strain of the rod element of length dx—is equal to

$$\varepsilon = \frac{(u + \mathrm{d}u) - u}{\mathrm{d}x},$$

that is

$$\frac{\mathrm{d}u}{\mathrm{d}x} = \varepsilon. \tag{1.3}$$

If strain is caused by acting stresses only, then according to the elasticity law $\varepsilon = \frac{1}{E}\sigma$. In the more general case

$$\varepsilon = \frac{1}{E}\sigma + e. \tag{1.4}$$

Strain $e = e(x)$ can occur as a result of thermal, magnetostrictive and piezo-electric effects when the rod is placed into a thermal, magnetic or electric field. Let us call $e = e(x)$ the strain from external action.

By virtue of the formulas (1.1) and (1.4)

$$\varepsilon = bQ + e, \tag{1.5}$$

where $b = \frac{1}{ES}$. We call $b = b(x)$ the distributed pliability or elasticity of a rod in tension.

As far as vibrations are concerned a rod is subjected to a distributed longitudinal load, which depends not only on coordinate x but also on time t: $p = p(x, t)$. Moreover, a rod can be placed in a nonstationary thermal, magnetic or electric field. Therefore, $e = e\,(x, t)$. Hence, the internal force $Q = Q\,(x, t)$ and displacement $u = u(x, t)$ will be functions not only of the coordinate but also of time.

To move from the static equations (1.2), (1.3), (1.5) to the equations of oscillations, distributed inertial forces should be included in the load acting on the rod.

They arise as a result of vibrations and are equal to $-\rho\ddot{u}$, where the dot denotes differentiation with respect to time. Derivatives with respect to coordinate x should also be replaced with partial derivatives. We obtain

$$-\frac{\partial Q}{\partial x} = h, \quad h = -\rho\ddot{u} + p, \qquad \frac{\partial u}{\partial x} = \varepsilon, \quad \varepsilon = bQ + e \tag{1.6}$$

or

$$-\frac{\partial Q}{\partial x} + \rho\ddot{u} - p = 0, \qquad \frac{\partial u}{\partial x} - bQ - e = 0. \tag{1.7}$$

Equations (1.7) describe the longitudinal vibrations of a rod. The first is the equation of dynamic equilibrium. The second is the law of elasticity, in which strains are expressed in terms of displacements. The basic unknowns of the problem are the internal force in a rod $Q = Q\,(x,t)$ and displacement $u = u(x,t)$.

Equations (1.7) should be augmented by boundary conditions, defined as the ends of a rod. Namely, on both ends of a rod at $x = 0, l$ at any moment in time t either force $Q = Q\,(x,t)$ or displacement $u = u(x,t)$ are considered to be known. Let us assume first that homogeneous boundary conditions take place at $x = 0, l$:

$$Q(x,t) = 0, \quad u(x,t) = 0, \tag{1.8}$$

if the end of a rod is free or fixed, respectively.

Two problems can be distinguished in oscillation theory:

(1) The initial value problem, in which displacements and velocities of all rod cross-sections are presumed to be known for a certain instant of time:

$$u(x,0) = u_0(x,0), \quad \dot{u}(x,0) = v_0(x,0), \quad 0 \le x \le l. \tag{1.9}$$

(2) The problem of periodic oscillations. In this case the periodic functions of time are the given external load $p(x,t) = p(x,t+T)$ and strain $e(x,t) = e(x,t+T)$, where T is the period of oscillations, $0 \le x \le l$. The periodicity conditions for displacements and velocities of rod motion are written as

$$u(x,0) = u(x,0+T), \quad \dot{u}(x,0) = \dot{u}(x,0+T), \quad 0 \le x \le l. \tag{1.10}$$

Equations (1.7), boundary conditions (1.8) and initial conditions (1.9) or periodicity conditions (1.10) completely describe the longitudinal vibrations of a rod of variable cross-section with a rectilinear axis.

Let us note the important particular case of harmonic oscillations when

$$p(x,t) = p(x)\cos(\lambda t + \alpha), \quad e(x,t) = e(x)\cos(\lambda t + \alpha). \tag{1.11}$$

It is here that force and displacement are also sought in the form of harmonic functions of time

$$\begin{aligned} Q(x,t) &= Q(x)\cos(\lambda t + \alpha), \\ u(x,t) &= u(x)\cos(\lambda t + \alpha), \end{aligned} \tag{1.12}$$

where $Q(x)$, $u(x)$ are the amplitudes of force and displacement, respectively. Moreover, $h(x,t) = h(x)\cos(\lambda t + \alpha)$, $\varepsilon(x,t) = \varepsilon(x)\cos(\lambda t + \alpha)$.

Formulas (1.6), (1.7), coupled with $\ddot{u}(x,t) = -\lambda^2 u(x)\cos(\lambda t + \alpha)$, allow the equations for amplitudes of force $Q = Q(x)$ and displacement $u = u(x)$ to be given as follows

$$\begin{aligned} -\frac{\mathrm{d}Q}{\mathrm{d}x} &= h, \quad h = \lambda^2 \rho u + p, \\ \frac{\mathrm{d}u}{\mathrm{d}x} &= \varepsilon, \quad \varepsilon = bQ + e \end{aligned} \tag{1.13}$$

or

$$\begin{aligned} -\frac{\mathrm{d}Q}{\mathrm{d}x} - \lambda^2 \rho u - p = 0, \\ \frac{\mathrm{d}u}{\mathrm{d}x} - bQ - e = 0. \end{aligned} \tag{1.14}$$

The desired functions (1.12) obviously satisfy the conditions (1.10) at the period $T = \frac{2\pi}{\lambda}$. The amplitude values of internal forces and displacements should, according to (1.8) and (1.12), satisfy the boundary conditions at $x = 0, l$:

$$\begin{aligned} Q = 0 \quad &\text{on the free end}, \\ u = 0 \quad &\text{on the fixed end}. \end{aligned} \tag{1.15}$$

1.2 Differential Equations of Longitudinal Vibrations of a Rod in the Operator Form

We apply the unified operator form of equations, to which the problem of vibrations of an elastic body can be reduced. In the case of longitudinal oscillations of a rod, let us introduce the following notation

$$\xi = Q, \quad \eta = u, \quad f = p, \tag{1.16}$$

$$D = -\frac{\partial}{\partial x}, \quad D^* = \frac{\partial}{\partial x}, \quad R = \rho, \quad B = b, \tag{1.17}$$

such that

$$D\xi = -\frac{\partial Q}{\partial x}, \quad D^*\eta = \frac{\partial u}{\partial x}, \quad R\eta = \rho u, \quad B\xi = bQ. \tag{1.18}$$

Let us impart such a form to Eqs. (1.6), (1.7)

$$\begin{aligned} D\xi &= h, \quad h = -R\ddot{\eta} + f, \\ D^*\eta &= \varepsilon, \quad \varepsilon = B\xi + e \end{aligned} \tag{1.19}$$

or

$$\begin{aligned} D\xi + R\ddot{\eta} - f &= 0, \\ D^*\eta - B\xi - e &= 0. \end{aligned} \tag{1.20}$$

where ξ characterizes the stress state; and η is the displacement of elastic body points. The form of the differential operators D and D^* depends only on the coordinate system selected and is linked with the dimension and shape of the elastic body considered, while the algebraic operators R and B define, respectively, the inertial and elastic properties of the body.

According to the homogeneous boundary conditions (1.8) at $x = 0, l$

$$\text{either } \xi = 0, \text{ or } \eta = 0, \tag{1.21}$$

such that

$$\xi\eta = 0. \tag{1.22}$$

Let us point out some important properties of the operators introduced.

By virtue of the formula of integration by parts

$$\int_0^l -\frac{\partial Q}{\partial x} u \mathrm{d}x = -Qu\Big|_{x=0}^{x=l} + \int_0^l Q\frac{\partial u}{\partial x}\mathrm{d}x,$$

such that

$$\int_0^l D\xi\eta \mathrm{d}x + \xi\eta\Big|_{x=0}^{x=l} = \int_0^l \xi D^*\eta \mathrm{d}x \tag{1.23}$$

The operators D and D^* connected with this equality contain boundary values of the functions on which equality (1.23) is performed. These operators are called conjugate or adjoint operators in the sense of Lagrange [1].

With homogeneous boundary conditions (1.21) as a consequence of (1.22)

$$\int_0^l D\xi\eta \mathrm{d}x = \int_0^l \xi D^* \eta \mathrm{d}x. \tag{1.24}$$

The operators D and D^* associated with this equality are called simply conjugate (adjoint). Note the physical meaning of (1.24). Using (1.19), rewrite (1.24) as

$$\frac{1}{2}\int_0^l h\eta \mathrm{d}x = \frac{1}{2}\int_0^l \xi\varepsilon \mathrm{d}x. \tag{1.25}$$

Hence, it is evident that the property of adjointness (conjugacy) of the operators D and D^* is equivalent to assertion of the equality of external load work and elastic strain energy.

The operators of inertia R and elasticity B are self-adjoint (self-conjugate) and positive

$$\int_0^l R\eta_1\eta_2 \mathrm{d}x = \int_0^l \eta_1 R\eta_2 \mathrm{d}x, \tag{1.26}$$

$$\int_0^l R\eta\eta \mathrm{d}x > 0 \quad \text{at} \quad \eta \neq 0, \tag{1.27}$$

since for any two displacements $u_1(x,t)$ and $u_2(x,t)$

$$\int_0^l (\rho u_1)u_2 \mathrm{d}x = \int_0^l u_1(\rho u_2)\mathrm{d}x \tag{1.28}$$

is performed and, moreover

$$\int_0^l (\rho u)u \mathrm{d}x > 0 \quad \text{at} \quad u(x,t) \neq 0 \tag{1.29}$$

since the distributed mass $\rho(x) > 0, \quad 0 \leq x \leq l$. Here and below, the signs = and $\neq$ in relation to functions mean identical equality and inequality, respectively.

Similarly

$$\int_0^l \xi_1 B\xi_2 \mathrm{d}x = \int_0^l B\xi_1 \xi_2 \mathrm{d}x, \tag{1.30}$$

$$\int_0^l \xi B\xi \mathrm{d}x > 0 \quad \text{at} \quad \xi \neq 0 \tag{1.31}$$

Really

$$\int_0^l Q_1(bQ_2)\mathrm{d}x = \int_0^l (bQ_1)Q_2\mathrm{d}x \tag{1.32}$$

and

$$\int_0^l Q(bQ)\mathrm{d}x > 0 \quad \text{at} \quad Q(x,t) \neq 0, \tag{1.33}$$

as the coefficient of distributed elasticity $b(x) > 0, \quad 0 \le x \le l.$

With harmonic vibrations, from Eqs. (1.19) it follows

$$\begin{aligned} D\xi &= h, \quad h = \lambda^2 R\eta + f, \\ D^*\eta &= \varepsilon, \quad \varepsilon = B\xi + e, \end{aligned} \tag{1.34}$$

or

$$\begin{aligned} &D\xi - \lambda^2 R\eta - f = 0, \\ &D^*\eta - B\xi - e = 0, \end{aligned} \tag{1.35}$$

where

$$D = -\frac{\mathrm{d}}{\mathrm{d}x}, \quad D^* = \frac{\mathrm{d}}{\mathrm{d}x}, \tag{1.36}$$

and $\xi = \xi(x), \quad \eta = \eta(x)$ are the amplitudes of forces and displacements.

In Eqs. (1.20) and (1.35), as well as later, it is not indicated whether the quantities in the equations are functions of both the coordinate and time or just the coordinate, as is clear from the problem conditions considered. Moreover, no distinction is made in the notation of operators (1.17) and (1.36) with partial or total derivatives with respect to the coordinate.

1.3 Differential Equations of Torsional Vibrations

Let us first assume that external static distributed torque $\mu(x)$ with respect to the axis x acts on a rod with a circular cross-section. Under its action, shearing stresses arise in each cross-section of the rod. The principal moment of these stresses will be called the “twisting moment” and denoted by $M(x)$. The equilibrium equation of the rod element of length dx has the form

$$(M + \mathrm{d}M) - M + \mu \mathrm{d}x = 0$$

or

$$-\frac{\mathrm{d}M}{\mathrm{d}x} = \mu. \tag{1.37}$$

Furthermore, denote by $\vartheta(x)$ the cross-section rotation angle about the axis x. The twisting of a rod is characterized by the quantity $\gamma = \frac{\mathrm{d}\vartheta}{\mathrm{d}x}$. By virtue of the law of elasticity

$$\gamma = \beta M + k, \tag{1.38}$$

where $\beta = \beta(x) = \frac{1}{GJ_p}$ is the elasticity of a rod under torsion; and $k = k(x)$ is the twisting of a rod caused by external impacts.

Moving on to the problem of torsional vibrations, replace the total derivatives with respect to x in Eqs. (1.37), (1.38) with the partial ones and represent the total distributed moment in the form

$$\mu = -j\ddot{\vartheta} + m, \tag{1.39}$$

where $m = m(x, t)$ is external distributed torque; and $j = j(x) = \rho_v J_p$ is the distributed moment of inertia of a rod.

We obtain

$$\begin{aligned} -\frac{\partial M}{\partial x} &= \mu, \quad \mu = -j\ddot{\vartheta} + m, \\ \frac{\partial \vartheta}{\partial x} &= \gamma, \quad \gamma = \beta M + k, \end{aligned} \tag{1.40}$$

or

$$\begin{aligned} -\frac{\partial M}{\partial x} + j\ddot{\vartheta} - m &= 0, \\ \frac{\partial \vartheta}{\partial x} - \beta M - k &= 0 \end{aligned} \tag{1.41}$$

Conditions at the ends of the rod are here as follows. At $x = 0, l$

$$M(x,t) = 0, \quad \vartheta(x,t) = 0,$$

if the end of the rod is free or fixed, respectively.

The initial conditions or periodicity conditions by time can be applied to the rotation angle $\vartheta = \vartheta(x,t)$ with the rotation angle changing velocity $\dot{\vartheta} = \dot{\vartheta}(x,t)$.

In the operator form with the notation

$$\begin{gathered} \xi = M, \ \ \eta = \vartheta, \ \ f = m, \ \ e = k, \ \ h = \mu, \ \ \varepsilon = \gamma, \\ D\xi = -\frac{\partial M}{\partial x}, \ \ D^{*}\eta = \frac{\partial \vartheta}{\partial x}, \ \ R = j, \ \ B = \beta \end{gathered} \tag{1.42}$$

the equations of torsional oscillations (1.40), (1.41) completely coincide with Eqs. (1.19), (1.20).

1.4 Differential Equations of Transverse Vibrations of a Rectilinear Rod

Let us start again with the static problem. We assume for definiteness that the external transverse distributed load $q(x)$ and external distributed moment $\mu(x)$ act on a rod, causing displacements of the rod only in the plane xy. With such a loading of a rod in its cross-section there arise tangential and bending stresses and, corresponding to them, shearing force $Q(x)$ and bending moment $M(x)$. Rod element equilibrium equations are composed of the condition of equality to zero of the sums of the forces and moments applied to this element

$$\begin{gathered} -Q + (Q + \mathrm{d}Q) + q\mathrm{d}x = 0, \\ -M + (M + \mathrm{d}M) - Q\mathrm{d}x + \mu\mathrm{d}x = 0. \end{gathered}$$

That is

$$\begin{gathered} -\frac{\mathrm{d}Q}{\mathrm{d}x} = q, \\ Q - \frac{\mathrm{d}M}{\mathrm{d}x} = \mu. \end{gathered} \tag{1.43}$$

In the absence of shear strain, the cross-sections of a rod during pure bending are rotated, staying flat and normal to its elastic line. Here $-\frac{\mathrm{d}u}{\mathrm{d}x} = \vartheta$, where $u = u(x)$ is the displacement of the point lying on the axis of the rod in the direction of the axis y; and $\vartheta = \vartheta(x)$ is the rotation angle of the cross-section of a rod.

At the same time, under the action of the shearing force the shear strain arises and, as a consequence, the additional angle of inclination $\gamma(x)$ of a rod's cross-section occurs: $-\frac{du}{dx}+\gamma=\vartheta$. By virtue of the law of elasticity $\gamma=bQ, b=\frac{\alpha}{GS}$, where α is a coefficient depending on the shape of the cross-section [2].

Moreover, the curvature of a rod $\kappa=\frac{d\vartheta}{dx}$ is proportional to the bending moment $\kappa=\beta M, \beta=\frac{1}{EJ}$. If there are additional changes in the cross-section slope angle and curvature of a rod, g and k, originating from the action of heat, magnetic and electric fields, then the elasticity equations can be written as follows

$$\begin{aligned} &\frac{du}{dx}+\vartheta=\gamma, \quad && \gamma=bQ+g,\\ &\frac{d\vartheta}{dx}+\kappa, && \kappa=\beta M+k. \end{aligned} \tag{1.44}$$

Let us use relations (1.43), (1.44) to compile the equations of transverse vibrations of a rod. Suppose that outward transverse load $p=p(x,t)$ and external distributed moment $m=m(x,t)$, depending both on the coordinate and time, act on a rod. In this case the desired functions of the coordinate and time will be $Q=Q(x,t), \quad M=M(x,t), \quad \vartheta=\vartheta(x,t), \quad u=u(x,t)$.

Total transverse load and distributed moment are composed, respectively, of external and inertial loads and moments:

$$\begin{aligned} &-\frac{\partial M}{\partial x}+j\ddot{\vartheta}-m=0,\\ &\frac{\partial \vartheta}{\partial x}-\beta M-k=0. \end{aligned} \tag{1.45}$$

where $j=j(x)=\rho_v J(x)$.

Substituting transverse load q and distributed moment μ from (1.45) into (1.43) and replacing the total derivatives in (1.43), (1.44) by partial ones, we get the equations of transverse vibrations of a rod:

$$\begin{aligned} -\frac{\partial Q}{\partial x}+\rho\ddot{u}-p=0, &\quad \frac{\partial u}{\partial x}+\vartheta-bQ-g=0,\\ Q-\frac{\partial M}{\partial x}+j\ddot{\vartheta}-m=0, &\quad \frac{\partial \vartheta}{\partial x}-\beta M-k=0. \end{aligned} \tag{1.46}$$

In the initial value problem for time moment $t=0$ are specified $u(x,0), \dot{u}(x,0), \vartheta(x,0), \dot{\vartheta}(x,0)$. In the problem of periodic oscillations the values of the quantities $u(x,t), \dot{u}(x,t), \vartheta(x,t), \dot{\vartheta}(x,t)$ are repeated through the time period T.

The boundary conditions at $x=0,l$

$$
\begin{array}{ll}
Q = 0, M = 0 & \text{on the free end} \\
Q = 0, \vartheta = 0 & \text{on the sliding sealing end} \\
M = 0, u = 0 & \text{on the simply or freely supported end} \\
u = 0, \vartheta = 0 & \text{on the built-in or fixed end}
\end{array}
\tag{1.47}
$$

In the case of harmonic oscillations, when

$$p(x,t) = p(x)\cos(\lambda t + \alpha), \quad m(x,t) = m(x)\cos(\lambda t + \alpha),$$

there will be

$$
\begin{aligned}
Q(x,t) &= Q(x)\cos(\lambda t + \alpha), \quad M(x,t) = M(x)\cos(\lambda t + \alpha), \\
u(x,t) &= u(x)\cos(\lambda t + \alpha), \quad \vartheta(x,t) = \vartheta(x)\cos(\lambda t + \alpha).
\end{aligned}
$$

In this case such equations follow from Eqs. (1.46) for the amplitudes of shearing force $Q = Q(x)$, bending moment $M = M(x)$, rotation angle of the cross-section $\vartheta = \vartheta(x)$ and displacements of rod axis points $u = u(x)$:

$$
\begin{aligned}
-\frac{\mathrm{d}Q}{\mathrm{d}x} - \lambda^2 \rho u - p = 0, &\quad \frac{\mathrm{d}u}{\mathrm{d}x} + \vartheta - bQ - g = 0, \\
Q - \frac{\mathrm{d}M}{\mathrm{d}x} - \lambda^2 j\vartheta - m = 0, &\quad \frac{\mathrm{d}\vartheta}{\mathrm{d}x} - \beta M - k = 0.
\end{aligned}
\tag{1.48}
$$

For a thin rod the distributed moment and shear strain can be neglected. This means that

$$j = 0, \quad m = 0, \quad b = 0, \quad g = 0. \tag{1.49}$$

Instead of Eqs. (1.46) we have the following

$$
\begin{aligned}
-\frac{\partial Q}{\partial x} + \rho \ddot{u} - p = 0, &\quad \frac{\partial u}{\partial x} + \vartheta = 0, \\
Q - \frac{\partial M}{\partial x} = 0, &\quad \frac{\partial \vartheta}{\partial x} - \beta M - k = 0.
\end{aligned}
\tag{1.50}
$$

Shearing force $Q = Q(x,t)$ and rotation angle $\vartheta = \vartheta(x,t)$ can be easily excluded from these equations. The system of two equations turns out as

$$
\begin{aligned}
&-\frac{\partial^2 M}{\partial x^2} + \rho \ddot{u} - p = 0, \\
&-\frac{\partial^2 u}{\partial x^2} - \beta M - k = 0.
\end{aligned}
\tag{1.51}
$$

The boundary conditions retain the form (1.47). However, by virtue of (1.50), there should be $Q = \frac{\partial M}{\partial x}$, $\vartheta = -\frac{\partial u}{\partial x}$.

1.5 Differential Equations of Transverse Vibrations of a Rod in the Operator Form

Let us write Eqs. (1.46), (1.48) in operator form. Let us also introduce the unknowns

$$\xi = (Q, M), \quad \eta = (u, \vartheta), \tag{1.52}$$

the differential operators

$$D = \begin{pmatrix} -\frac{\partial}{\partial x} & 0 \\ 1 & -\frac{\partial}{\partial x} \end{pmatrix}, \quad D^* = \begin{pmatrix} \frac{\partial}{\partial x} & 1 \\ 0 & \frac{\partial}{\partial x} \end{pmatrix}, \tag{1.53}$$

and the algebraic operators of inertia and elasticity

$$R = \begin{pmatrix} \rho & 0 \\ 0 & j \end{pmatrix}, \quad B = \begin{pmatrix} b & 0 \\ 0 & \beta \end{pmatrix}. \tag{1.54}$$

The variable ξ characterizes the stress state in a rod, while the variable η determines displacements of the points of a rod. We will call ξ and η, accordingly, the generalized internal force and generalized displacement, or simply the force and displacement. Hereafter quantities ξ and η are vectors with the projections Q, M and u, ϑ, making no distinction in the notation between a row vector and a column vector.

Let us also introduce the generalized external load

$$f = (p, m) \tag{1.55}$$

and generalized strain from external impacts

$$e = (g, k). \tag{1.56}$$

In the notation introduced in (1.52)–(1.56) the differential equations (1.46), (1.48) are also written in the form of equations with differential operators (1.20), (1.35).

On the basis of the simple homogeneous boundary conditions (1.47) at $x = 0, l$

$$\xi\eta = Qu + M\vartheta = 0. \tag{1.57}$$

Note the properties of operators (1.53), (1.54). The operators D and D^* are conjugate in the sense of Lagrange. That is, equality (1.23) is valid. Hence,

$$D\xi = \begin{pmatrix} -\frac{\partial}{\partial x} & 0 \\ 1 & -\frac{\partial}{\partial x} \end{pmatrix} \begin{pmatrix} Q \\ M \end{pmatrix} = \begin{pmatrix} -\frac{\partial Q}{\partial x} \\ Q - \frac{\partial M}{\partial x} \end{pmatrix},$$

$$D^*\eta = \begin{pmatrix} \frac{\partial}{\partial x} & 1 \\ 0 & \frac{\partial}{\partial x} \end{pmatrix} \begin{pmatrix} u \\ \vartheta \end{pmatrix} = \begin{pmatrix} \frac{\partial u}{\partial x} + \vartheta \\ \frac{\partial \vartheta}{\partial x} \end{pmatrix}$$

and according to the rule of integration by parts

$$\begin{aligned} \int_0^l D\xi\eta \mathrm{d}x &= \int_0^l \left(-\frac{\partial Q}{\partial x} u + \left(Q - \frac{\partial M}{\partial x} \right) \vartheta \right) \mathrm{d}x = (-Qu - M\vartheta)\Big|_{x=0}^{x=l} \\ &+ \int_0^l \left(Q\left(\frac{\partial u}{\partial x} + \vartheta \right) + M\frac{\partial \vartheta}{\partial x} \right) \mathrm{d}x = -\xi\eta\Big|_{x=0}^{x=l} + \int_0^l \xi D^*\eta \mathrm{d}x. \end{aligned} \tag{1.58}$$

By virtue of (1.57), the operators D and D^* are merely adjoint (conjugate) (i.e. equality (1.24) holds).

The operator of inertia R is self-adjoint (self-conjugate). We have

$$R\eta = \begin{pmatrix} \rho & 0 \\ 0 & j \end{pmatrix} \begin{pmatrix} u \\ \vartheta \end{pmatrix} = \begin{pmatrix} \rho u \\ j\vartheta \end{pmatrix},$$

such that

$$\int_0^l R\eta_1\eta_2 \mathrm{d}x = \int_0^l (\rho u_1 u_2 + j\vartheta_1\vartheta_2)\mathrm{d}x = \int_0^l (u_1 \rho u_2 + \vartheta_1 j\vartheta_2)\mathrm{d}x = \int_0^l \eta_1 R\eta_2 \mathrm{d}x.$$

The operator of inertia R, moreover, is positive

$$\int_0^l R\eta\eta \mathrm{d}x = \int_0^l (\rho uu + j\vartheta\vartheta)\mathrm{d}x > 0, \quad \text{if} \quad \eta \neq 0.$$

The operator of elasticity B is also self-adjoint (self-conjugate) and positive

$$B\xi = \begin{pmatrix} b & 0 \\ 0 & \beta \end{pmatrix} \begin{pmatrix} Q \\ M \end{pmatrix} = \begin{pmatrix} bQ \\ \beta M \end{pmatrix},$$

$$\int_0^l B\xi_1\xi_2 \mathrm{d}x = \int_0^l (bQ_1Q_2 + \beta M_1M_2)\mathrm{d}x$$
$$= \int_0^l (Q_1 bQ_2 + M_1\beta M_2)dx = \int_0^l \xi_1 B\xi_2 \mathrm{d}x$$

and

$$\int_0^l B\xi\xi \mathrm{d}x = \int_0^l (bQQ + \beta MM)\mathrm{d}x > 0, \text{ if } \xi \neq 0.$$

In what follows, we will need the concepts of generalized total load and generalized total strain, which we simply call the "load" and "strain"

$$h = (q, \mu), \quad \varepsilon = (\gamma, \kappa). \tag{1.59}$$

According to (1.46)

$$D\xi = h, \quad D^*\eta = \varepsilon. \tag{1.60}$$

On the basis of (1.44), (1.45)

$$h = -R\ddot{\eta} + f, \quad \varepsilon = B\xi + e, \tag{1.61}$$

and with harmonic oscillations

$$h = \lambda^2 R\eta + f, \quad \varepsilon = B\xi + e. \tag{1.62}$$

It is obvious that (1.60), (1.61) are equivalent to Eqs. (1.20), while (1.60), (1.62) are equivalent to Eqs. (1.35).

In the case of a thin rod let us introduce the notation

$$\xi = M, \quad \eta = u, \tag{1.63}$$

the differential operators

$$D = -\frac{\partial^2}{\partial x^2}, \quad D^* = -\frac{\partial^2}{\partial x^2}, \tag{1.64}$$

the operators of inertia and elasticity

$$R = \rho, \quad B = \beta, \tag{1.65}$$

the transverse distributed load on a rod

$$f = p, \tag{1.66}$$

and the given strain

$$e = k. \tag{1.67}$$

Then, Eqs. (1.51) will be written in the form (1.20). Under the conditions (1.47) the relations (1.24) hold true. The operators R and B of (1.65) are obviously self-adjoint and positive.

1.6 Joint Longitudinal, Torsional and Transverse Vibrations of a Rod

The operator form of (1.20) is applicable also in the case of joint longitudinal, torsional and transverse vibrations of a rectilinear rod. Hence, in equations of the form (1.20) there will be

$$f = (p, m), \quad p = (p^x, p^y, p^z), \quad m = (m^x, m^y, m^z),$$

where p^x, p^y, p^z and m^x, m^y, m^z are the projections of external distributed load and external distributed moment on the axes x, y, z.

Transverse oscillations are considered in two planes xy, xz. Analogously, let us introduce the vector force $Q = (Q^x, Q^y, Q^z)$ and moment $M = (M^x, M^y, M^z)$ in an arbitrary cross-section of a rod, the vector displacement and rotation angle $u = (u_x, u_y, u_z)$, $\vartheta = (\vartheta_x, \vartheta_y, \vartheta_z)$ and the generalized force and displacement $\xi = (Q, M)$, $\eta = (u, \vartheta)$. The operators have the form

$$D = \begin{Bmatrix} -\frac{\partial}{\partial x} & 0 & 0 & 0 & 0 & 0 \\ 0 & -\frac{\partial}{\partial x} & 0 & 0 & 0 & 0 \\ 0 & 0 & -\frac{\partial}{\partial x} & 0 & 0 & 0 \\ 0 & 0 & 0 & -\frac{\partial}{\partial x} & 0 & 0 \\ 0 & 0 & 1 & 0 & -\frac{\partial}{\partial x} & 0 \\ 0 & 1 & 0 & 0 & 0 & -\frac{\partial}{\partial x} \end{Bmatrix},$$

$$D^* = \begin{Bmatrix} \frac{\partial}{\partial x} & 0 & 0 & 0 & 0 & 0 \\ 0 & \frac{\partial}{\partial x} & 0 & 0 & 0 & 1 \\ 0 & 0 & \frac{\partial}{\partial x} & 0 & 1 & 0 \\ 0 & 0 & 0 & \frac{\partial}{\partial x} & 0 & 0 \\ 0 & 0 & 0 & 0 & \frac{\partial}{\partial x} & 0 \\ 0 & 0 & 0 & 0 & 0 & \frac{\partial}{\partial x} \end{Bmatrix},$$

$$R = \mathrm{diag}(\rho, \rho, \rho, j_x, j_y, j_z),$$
$$B = \mathrm{diag}(b_x, b_y, b_z, \beta_x, \beta_y, \beta_z),$$

where j_x, j_y, j_z are the distributed moments of inertia of a rod at angular displacement relative to the axes x, y, z, b_x, b_y, b_z are the coefficients of distributed elasticity in tension and shear; and $\beta_x, \beta_y, \beta_z$ are the same coefficients at torsion and bending.

The properties of the operators (1.23), (1.26), (1.27), (1.30), (1.31) remain true. So, for example,

$$\int_0^l D\xi\eta \mathrm{d}x = \int_0^l \left(-\frac{\partial Q^x}{\partial x} u_x - \frac{\partial Q^y}{\partial x} u_y - \frac{\partial Q^z}{\partial x} u_z - \frac{\partial M^x}{\partial x} \vartheta_x \right.$$
$$\left. + \left(Q^z - \frac{\partial M^y}{\partial x} \right) \vartheta_y + \left(Q^y - \frac{\partial M^z}{\partial x} \right) \vartheta_z \right) \mathrm{d}x$$
$$= \int_0^l \left(Q^x \frac{\partial u_x}{\partial x} + Q^y \left(\frac{\partial u_y}{\partial x} + \vartheta_z \right) + Q^z \left(\frac{\partial u_z}{\partial x} + \vartheta_y \right) + M^x \frac{\partial \vartheta_x}{\partial x} \right.$$
$$\left. + M^y \frac{\partial \vartheta_y}{\partial x} + M^z \frac{\partial \vartheta_z}{\partial x} \right) \mathrm{d}x - \left(Q^x u_x + Q^y u_y + Q^z u_z + M^x \vartheta_x \right.$$
$$\left. + M^y \vartheta_y + M^z \vartheta_z \right) \Big|_{x=0}^{x=l} = \int_0^l \xi D^* \eta \mathrm{d}x - \xi\eta \Big|_{x=0}^{x=l}.$$

The more general case of a curvilinear rod is examined in the Appendix.

1.7 Differential Equations in Displacements and Forces

Let us introduce the inverse of the inertia and elasticity operators and denote them, respectively, by R^{-1} and B^{-1}. The inverse operators are found from the equalities

$$\eta = R^{-1} R \eta, \quad \xi = B^{-1} B \xi. \tag{1.68}$$

It is evident that in the problem of longitudinal vibrations

$$R^{-1} = \frac{1}{\rho}, \quad B^{-1} = \frac{1}{b}. \tag{1.69}$$

In the case of transverse vibrations

$$R^{-1} = \begin{pmatrix} \frac{1}{\rho} & 0 \\ 0 & \frac{1}{j} \end{pmatrix}, \quad B^{-1} = \begin{pmatrix} \frac{1}{b} & 0 \\ 0 & \frac{1}{\beta} \end{pmatrix}. \tag{1.70}$$

Let us move from the system of equations in both forces and displacements of the form (1.20), (1.35) to equations with respect to displacements only. For this purpose we multiply the second equation of (1.20) by the operator B^{-1}

$$B^{-1}D^*\eta - B^{-1}B\xi - B^{-1}e = 0.$$

We have, according to (1.68)

$$\xi = B^{-1}D^*\eta - B^{-1}e. \tag{1.71}$$

Substituting the force ξ from (1.71) into the first Eq. (1.20), we obtain:

$$DB^{-1}D^*\eta + R\ddot{\eta} - (f + DB^{-1}e) = 0. \tag{1.72}$$

To simplify notation we denote the inhomogeneous term of Eq. (1.72) by f, keeping in mind that the inhomogeneous term has the more complicated form $f + DB^{-1}e$ when strain from external action e is present.

Introducing the operator

$$N = DB^{-1}D^*, \tag{1.73}$$

we write the equation in displacements (1.72) as follows

$$N\eta + R\ddot{\eta} - f = 0. \tag{1.74}$$

The equation relative to the amplitude of harmonic oscillations has the form

$$N\eta - \lambda^2 R\eta - f = 0. \tag{1.75}$$

The record of equations in the operator form is compact and convenient in many ways. Such a record is unified, as we have seen, in the cases of longitudinal, torsional and transverse vibrations of a rod. The form of the operator equations as well as general properties of the operators are preserved when considering oscillations of more complex elastic bodies.

At the same time, it is easy if necessary to switch from operator to differential equations, which is usually more cumbersome. We shall do this using the example of transverse vibrations of a rod.

We have

$$D^*\eta = \begin{pmatrix} \frac{\partial}{\partial x} & 1 \\ 1 & \frac{\partial}{\partial x} \end{pmatrix}\begin{pmatrix} u \\ \vartheta \end{pmatrix} = \begin{pmatrix} \frac{\partial u}{\partial x} + \vartheta \\ \frac{\partial \vartheta}{\partial x} \end{pmatrix}.$$

Further,

$$B^{-1}D^*\eta = \begin{pmatrix} \frac{1}{b} & 0 \\ 0 & \frac{1}{\beta} \end{pmatrix}\begin{pmatrix} \frac{\partial u}{\partial x} + \vartheta \\ \frac{\partial \vartheta}{\partial x} \end{pmatrix} = \begin{pmatrix} \frac{1}{b}\left(\frac{\partial u}{\partial x} + \vartheta\right) \\ \frac{1}{\beta}\frac{\partial \vartheta}{\partial x} \end{pmatrix}$$

and

$$\begin{aligned} N\eta = DB^{-1}D^*\eta &= \begin{pmatrix} -\frac{\partial}{\partial x} & 0 \\ 1 & -\frac{\partial}{\partial x} \end{pmatrix}\begin{pmatrix} \frac{1}{b}\left(\frac{\partial u}{\partial x} + \vartheta\right) \\ \frac{1}{\beta}\frac{\partial \vartheta}{\partial x} \end{pmatrix} \\ &= \begin{pmatrix} -\frac{\partial}{\partial x}\left(\frac{1}{b}\left(\frac{\partial u}{\partial x} + \vartheta\right)\right) \\ \frac{1}{b}\left(\frac{\partial u}{\partial x} + \vartheta\right) - \frac{\partial}{\partial x}\left(\frac{1}{\beta}\frac{\partial \vartheta}{\partial x}\right) \end{pmatrix}. \end{aligned} \tag{1.76}$$

Finally,

$$R\ddot{\eta} = \begin{pmatrix} \rho & 0 \\ 0 & j \end{pmatrix}\begin{pmatrix} \ddot{u} \\ \ddot{\vartheta} \end{pmatrix} = \begin{pmatrix} \rho\ddot{u} \\ j\ddot{\vartheta} \end{pmatrix}, \quad f = \begin{pmatrix} p \\ m \end{pmatrix}. \tag{1.77}$$

Thus, the operator Eq. (1.74) turns out to be equivalent to the system of differential equations

$$\begin{aligned} -\frac{\partial}{\partial x}\left(\frac{1}{b}\left(\frac{\partial u}{\partial x} + \vartheta\right)\right) + \rho\ddot{u} - p = 0, \\ \frac{1}{b}\left(\frac{\partial u}{\partial x} + \vartheta\right) - \frac{\partial}{\partial x}\left(\frac{1}{\beta}\frac{\partial \vartheta}{\partial x}\right) + j\ddot{\vartheta} - m = 0. \end{aligned} \tag{1.78}$$

In the case of a thin rod with the notation (1.63)–(1.66) and $e = 0$, Eq. (1.74) takes the form

$$\frac{\partial^2}{\partial x^2}\left(EJ\frac{\partial^2 u}{\partial x^2}\right) + \rho\ddot{u} - p = 0, \tag{1.79}$$

where such a substitution has been done $\beta = \frac{1}{EJ}$.

Let us now prove that the operator N of (1.74) is self-adjoint and positive. First, it is necessary to point out that the inverse operator B^{-1} is also self-adjoint and positive. This is a consequence of the same properties of the operator B, and can be verified directly as well:

$$\int_0^l B^{-1}\varepsilon_1\varepsilon_2 \mathrm{d}x = \int_0^l \varepsilon_1 B^{-1}\varepsilon_2 \mathrm{d}x,$$
$$\int_0^l B^{-1}\varepsilon\varepsilon \mathrm{d}x > 0, \text{ if } \varepsilon \neq 0. \tag{1.80}$$

Referring to (1.73), we have

$$\int_0^l N\eta_1\eta_2 \mathrm{d}x = \int_0^l DB^{-1}D^*\eta_1\eta_2 \mathrm{d}x,$$

so that, using the adjointness property of the operators D and D^*, as well as the self-adjointness property of the operator B^{-1}, we can write

$$\int_0^l N\eta_1\eta_2 \mathrm{d}x = \int_0^l DB^{-1}D^*\eta_1\eta_2 \mathrm{d}x = \int_0^l B^{-1}D^*\eta_1 D^*\eta_2 \mathrm{d}x$$
$$= \int_0^l D^*\eta_1 B^{-1}D^*\eta_2 \mathrm{d}x = \int_0^l \eta_1 DB^{-1}D^*\eta_2 \mathrm{d}x = \int_0^l \eta_1 N\eta_2 \mathrm{d}x, \tag{1.81}$$

which proves the self-adjointness of the operator N.

The operator N is nonnegative; that is

$$\int_0^l N\eta\eta \mathrm{d}x \geq 0 \tag{1.82}$$

Hence,

$$\int_0^l N\eta\eta \mathrm{d}x = \int_0^l D^*\eta B^{-1}D^*\eta \mathrm{d}x = \int_0^l \varepsilon B^{-1}\varepsilon \mathrm{d}x \geq 0.$$

There are two possible cases at this point.

(1) Under the fastening conditions given the displacements of a rigid body are $\eta(x,t) \neq 0$, while the strain is $\varepsilon(x,t) = 0$
and

$$\int_0^l N\eta\eta\mathrm{d}x = \int_0^l \varepsilon B^{-1}\varepsilon\mathrm{d}x = 0. \tag{1.83}$$

(2) When $\varepsilon(x, t) \neq 0$, the operator N on the respective displacements is positive:

$$\int_0^l N\eta\eta\mathrm{d}x = \int_0^l \varepsilon B^{-1}\varepsilon\mathrm{d}x > 0. \tag{1.84}$$

Similarly, it is possible to move from the system of equations in both forces and displacements (1.20) to equations with respect to forces only. To do this, multiply the first of Eqs. (1.20) by R^{-1}, while differentiating the second one twice with respect to time

$$R^{-1}D\xi + R^{-1}R\ddot{\eta} - R^{-1}f = 0,$$
$$D^*\ddot{\eta} - B\ddot{\xi} - \ddot{e} = 0.$$

Hence, it follows that

$$D^*R^{-1}D\xi + B\ddot{\xi} - D^*R^{-1}f + \ddot{e} = 0. \tag{1.85}$$

1.8 Integral Equations of Longitudinal and Torsional Vibrations

Let us return to longitudinal vibrations of a rectilinear rod. First, let us transform the equations of statics. Having integrated (1.2), (1.3), we obtain

$$Q(x) = Q(0) - \int_0^x h\mathrm{d}s, \quad u(x) = u(0) + \int_0^x \varepsilon\mathrm{d}s, \quad 0 \le x \le l, \tag{1.86}$$

where $Q(0)$ and $u(0)$ are force and displacement, respectively, at the left end of a rod. The concrete form of integral equations depends on boundary conditions. Our reasoning uses the example of a rod with boundary conditions

$$Q(0) = 0, \quad u(l) = 0. \tag{1.87}$$

Such conditions (i.e. one given on the force and the other on displacement) are very convenient to transit from differential equations to integral ones. By virtue of (1.87) we have

$$Q(x) = -\int_0^x h\mathrm{d}s, \quad u(l) = u(0) + \int_0^l \varepsilon\mathrm{d}s = 0,$$

such that

$$u(0) = -\int_0^l \varepsilon\mathrm{d}s, \quad u(x) = -\int_0^l \varepsilon\mathrm{d}s + \int_0^x \varepsilon\mathrm{d}s = -\int_x^l \varepsilon\mathrm{d}s.$$

The derived relations

$$Q(x) = -\int_0^x h(s)\mathrm{d}s, \quad u(x) = -\int_x^l \varepsilon(s)\mathrm{d}s$$

can be put in an enhanced way

$$Q(x) = \int_0^l K(x,s)h(s)ds, u(x) = \int_0^l K(s,x)\varepsilon(s)ds. \tag{1.88}$$

This equation introduces the kernel (the function of two variables x and s)

$$K(x,s) = \begin{cases} -1, & 0 \le s \le x, \\ 0, & x < s \le l. \end{cases} \tag{1.89}$$

The kernel $K(s,x)$ is obtained by mutually interchanging x on s and s on x and is called the “kernel,” “adjoint” or “conjugate” of $K(x,s)$

$$K(s,x) = \begin{cases} 0, & 0 \le s < x, \\ -1, & x \le s \le l. \end{cases}$$

In the problem of vibrations $Q = Q(x,t), u = u(x,t)$ and, according to (1.6)

$$h(x,t) = -\rho(x)\ddot{u}(x,t) + p(x,t),$$
$$\varepsilon(x,t) = b(x)Q(x,t) + e(x,t).$$

Hence, by virtue of (1.88) we arrive at the system of integral equations with respect to the internal force $Q = Q(x,t)$ and displacement $u = u(x,t)$

$$\begin{aligned} & Q(x,t) + \int_0^l K(x,s)\rho(s)\ddot{u}(s,t)\mathrm{d}s - \int_0^l K(x,s)p(s,t)\mathrm{d}s = 0, \\ & u(x,t) - \int_0^l K(s,x)b(s)Q(s,t)\mathrm{d}s - \int_0^l K(s,x)e(s,t)\mathrm{d}s = 0. \end{aligned} \tag{1.90}$$

The integral equations for the amplitudes of harmonic oscillations $Q = Q(x)$ and $u = u(x)$ are written as

$$\begin{aligned} & Q(x) - \lambda^2 \int_0^l K(x,s)\rho(s)u(s)\mathrm{d}s - \int_0^l K(x,s)p(s)\mathrm{d}s = 0, \\ & u(x) - \int_0^l K(s,x)b(s)Q(s)\mathrm{d}s - \int_0^l K(s,x)e(s)\mathrm{d}s = 0. \end{aligned} \tag{1.91}$$

We point out the characteristic properties of the kernel $K(x,s)$ having the form (1.89). Obviously

$$\int_0^l \int_0^l |K(x,s)|^2 \mathrm{d}x\mathrm{d}s < \infty, \tag{1.92}$$

$$\int_0^l \left(\int_0^l K(x,s)h(s)\mathrm{d}s \right) \varepsilon(x)\mathrm{d}x = \int_0^l h(s) \left(\int_0^l K(x,s)\varepsilon(x)\mathrm{d}x \right) \mathrm{d}s. \tag{1.93}$$

The kernel satisfying condition (1.92) is called the Fredholm kernel, while formulas (1.91) represent the Fredholm integral equations [3, 4].

Let us proceed to the operator record of integral equations. Formulas (1.88) are written more concisely as follows

$$\xi = Kh, \quad \eta = K^*\varepsilon. \tag{1.94}$$

Here we have returned to the notation of forces and displacements (1.17).

The operator K, defined by the integral operation (1.88) in which $K(x,s)$ is the Fredholm kernel, is called the Fredholm integral operator.

Equality (1.93) takes the form

$$\int_0^l Kh\varepsilon\mathrm{d}x = \int_0^l hK^*\varepsilon\mathrm{d}x. \tag{1.95}$$

This condition of adjointness of the operators K and K^* is a consequence of adjointness of the kernels $K(x,s)$ and $K(s,x)$.

Comparing formulas (1.94) and (1.19), we can write that

$$\xi = KD\xi, \quad \eta = K^*D^*\eta, \tag{1.96}$$

as well as

$$h = DKh, \quad \varepsilon = D^*K^*\varepsilon. \tag{1.97}$$

In other words, the integral operators K and K^* and the differential operators D and D^* are simply mutually inverse

$$K = D^{-1},\ K^* = (D^*)^{-1},\ D = K^{-1},\ D^* = (K^*)^{-1}. \tag{1.98}$$

The equations with integral operators, replacing the integral equations (1.90) and (1.91), appear as follows

$$\begin{aligned} &\xi + KR\ddot{\eta} - Kf = 0, \\ &\eta - K^*B\xi - K^*e = 0 \end{aligned} \tag{1.99}$$

and

$$\begin{aligned} &\xi - \lambda^2 KR\eta - Kf = 0, \\ &\eta - K^*B\xi - K^*e = 0. \end{aligned} \tag{1.100}$$

Note that Eqs. (1.99) are immediately obtained by multiplying the first and second Eqs. (1.20), respectively, by K and K^* on the basis of equalities (1.96). In like manner the transition from (1.35) to (1.100) can be carried out.

The equations of torsional vibrations, as we have seen, are similar to the equations of longitudinal vibrations, and therefore we will not dwell on them.

1.9 Integral Equations of Transverse Vibrations of a Rod

Let us integrate Eqs. (1.43):

$$\begin{aligned} &Q(x) = Q(0) - \int_0^x q\mathrm{d}s, \\ &M(x) = M(0) + \int_0^x Q\mathrm{d}s - \int_0^x \mu\mathrm{d}s. \end{aligned} \tag{1.101}$$

For simplicity, let us consider again a rod with one free end and another fixed (clamped or built-in) one. The boundary conditions are here as follows:

$$Q(0) = 0, \quad M(0) = 0, \quad u(l) = 0, \quad \vartheta(l) = 0. \tag{1.102}$$

Substituting $Q(x)$ from the first equation into the second one and taking into account (1.102) we get

$$\begin{aligned} Q(x) &= -\int_0^x q(s)\mathrm{d}s, \\ M(x) &= -\int_0^x \mathrm{d}s \int_0^s q(r)\mathrm{d}r - \int_0^x \mu(s)\mathrm{d}s \\ &= \int_0^x (s-x)q(s)\mathrm{d}s - \int_0^x \mu(s)\mathrm{d}s, \end{aligned} \tag{1.103}$$

since by virtue of the formula of integration by parts

$$\begin{aligned} \int_0^x \mathrm{d}s \int_0^s q(r)\mathrm{d}r &= s\int_0^s q(r)\mathrm{d}r\Big|_{s=0}^{s=x} - \int_0^x sq(s)\mathrm{d}s \\ &= \int_0^x (x-s)q(s)\mathrm{d}s. \end{aligned}$$

Let us write (1.103) in another way

$$\begin{aligned} Q(x) &= \int_0^l K_{11}(x,s)q(s)\mathrm{d}s + \int_0^l K_{12}(x,s)\mu(s)\mathrm{d}s, \\ M(x) &= \int_0^l K_{21}(x,s)q(s)\mathrm{d}s + \int_0^l K_{22}(x,s)\mu(s)\mathrm{d}s, \end{aligned} \tag{1.104}$$

introducing the kernels

$$K_{11}(x,s) = \begin{cases} -1, & 0 \le s \le x, \\ 0, & x < s \le l, \end{cases} \qquad K_{12}(x,s) = 0,$$

$$K_{21}(x,s) = \begin{cases} s-x, & 0 \le s \le x, \\ 0, & x < s \le l, \end{cases} \qquad K_{22}(x,s) = K_{11}(x,s).$$

In operator form, equalities (1.104) are as follows

$$\begin{aligned} Q &= K_{11}q + K_{12}\mu, \\ M &= K_{21}q + K_{22}\mu. \end{aligned} \tag{1.105}$$

Let us denote, as before, $\xi = (Q, M), h = (q, \mu)$, introduce the operator

$$K = \begin{pmatrix} K_{11} & K_{12} \\ K_{21} & K_{22} \end{pmatrix}, \tag{1.106}$$

and impart such a form to relations (1.105):

$$\xi = Kh. \tag{1.107}$$

Similarly, having integrated (1.44) at $u(l) = 0, \vartheta(l) = 0$, we can verify that

$$\eta = K^*\varepsilon. \tag{1.108}$$

Here

$$K^* = \begin{pmatrix} K_{11}^* & K_{21}^* \\ K_{12}^* & K_{22}^* \end{pmatrix}.$$

The adjoint (conjugate) kernels $K_{11}(s,x), \ldots, K_{22}(s,x)$ correspond to the adjoint (conjugate) operators $K_{11}^*, \ldots, K_{22}^*$.

Substituting load h and strain ε from (1.61) into (1.107), (1.108), we arrive at the system of equations with integral operators, which describes transverse vibrations of a rod of the form (1.99) and harmonic oscillations of the form (1.100).

The integral operators K and K^* here are also adjoint

$$\begin{aligned} \int_0^l Kh\varepsilon \mathrm{d}x &= \int_0^l ((K_{11}q + K_{12}\mu)\gamma + (K_{21}q + K_{22}\mu)\kappa)\mathrm{d}x \\ &= \int_0^l \left(q(K_{11}^*\gamma + K_{21}^*\kappa) + \mu(K_{12}^*\gamma + K_{22}^*\kappa)\right)\mathrm{d}x = \int_0^l hK^*\varepsilon \mathrm{d}x. \end{aligned}$$

Comparing formulas (1.107), (1.108) with (1.60), we see that the integral operators K and K^* are mutually inverse of the differential operators D and D^* , respectively.

For a thin rod in expressions (1.107), (1.108)

$$\begin{aligned}&\xi = M, \eta = u, R = \rho, B = \beta, f = p,\ e = k,\\&h = -\rho\ddot{u} + p, \varepsilon = \beta M + k,\\&K = K_{21}, K^{*} = K_{21}^{*}.\end{aligned}$$

Thus, it is shown that equations with both differential and integral operators are of like form in the cases of longitudinal, torsional and transverse vibrations of a rectilinear rod. Therefore, further arguments can be carried out using the general common operator form of the equations, while specific types of vibrations are considered for illustration.

Note that in calculations it is not necessary to constitute the expressions for kernels of the type (1.104); it is more convenient to use the representation of integral operators by integrals with a variable limit as, for example, in formula (1.103).

1.10 Equations in Displacements with Integral Operators

The equations in displacements with integral operators are obtained directly from the system of Eqs. (1.99) by substituting the force ξ from the first equation into the second one

$$\eta + K^{*}BKR\ddot{\eta} - K^{*}BKf - K^{*}e = 0$$

or

$$\eta + AR\ddot{\eta} - \varphi = 0, \tag{1.109}$$

where such an operator has been introduced

$$A = K^{*}BK \tag{1.110}$$

and has been denoted

$$Af + K^{*}e = \varphi. \tag{1.111}$$

The equation relative to the displacement amplitude at harmonic vibrations is of the form

$$\eta - \lambda^{2}AR\eta - \varphi = 0. \tag{1.112}$$

The introduced operator A is the Fredholm operator. Thus, at longitudinal vibrations

$$Af = K^*BKf = \int_0^l \left(K(r,x)b(r) \int_0^l K(r,s)f(s)\mathrm{d}s \right) \mathrm{d}r$$
$$= \int_0^l \left(\int_0^l K(r,x)b(r)K(r,s)\mathrm{d}r \right) f(s)\mathrm{d}s = \int_0^l A(x,s)f(s)\mathrm{d}s, \tag{1.113}$$

where the kernel is

$$A(x,s) = \int_0^l K(r,x)b(r)K(r,s)\mathrm{d}r.$$

The integral operator A of (1.110) and differential operator N of (1.73) are mutually inverse

$$AN\eta = K^*BKDB^{-1}D^*\eta = K^*BB^{-1}D^*\eta = K^*D^*\eta = \eta, \tag{1.114}$$

since, according to (1.68) and (1.98), K and D, B and B^{-1}, K^* and D^* are mutually inverse. Similarly

$$NAh = h \tag{1.115}$$

The operator A is self-adjoint and positive, because the inverse operator N only possesses these properties. Direct verification also gives

$$\int_0^l Ah_1h_2\mathrm{d}x = \int_0^l K^*BKh_1h_2\mathrm{d}x = \int_0^l BKh_1Kh_2\mathrm{d}x$$
$$= \int_0^l Kh_1BKh_2\mathrm{d}x = \int_0^l h_1K^*BKh_2\mathrm{d}x = \int_0^l h_1Ah_2\mathrm{d}x,$$
$$= \int_0^l Ahh\mathrm{d}x = \int_0^l K^*BKhh\mathrm{d}x = \int_0^l BKhKh\mathrm{d}x > 0, \text{ if } h \neq 0.$$

Let us clarify the physical meaning of the operator A. In the static problem, when $\ddot{\eta} = 0$ and strain as a result of external impact is absent ($e = 0$), displacement, according to (1.109), (1.111), is

$$\eta = \varphi = Af. \tag{1.116}$$

Hence, A is the operator by means of which the static displacement φ is calculated by the given external load f.

Let us pay attention to an important circumstance. It is necessary to distinguish two cases.

(1) The fastening conditions of a rod are such that they do not allow displacements of a rod as an absolutely rigid body—so-called “hard” movements.
(2) The fastening conditions of a rod permit one or several linearly independent rigid displacements.

Overall, there can be no more than six hard movements because of the number of possible displacements that are both translational and rotational.

It is essential that the abovementioned expressions for the integral operators K, K^* and A have been derived using the fastening conditions of a rod that do not allow rigid movement. The more complicated case, when hard movements are possible, is considered in Sect. 1.12.

1.11 Converting the Equations with Differential and Integral Operators to the Classical Form

Let us impart the classic form of notation to Eqs. (1.19), (1.20) and (1.74), which is acceptable in the theory of linear equations with self-adjoint operators.

Suppose, in accordance with reality, that a rod has no weightless or absolutely rigid portions. In the case of longitudinal vibrations

$$\rho_{max} \geq \rho(x) \geq \rho_{min} > 0 \tag{1.117}$$

and

$$b_{max} \geq b(x) \geq b_{min} > 0, 0 \leq x \leq l.$$

Hence, the operators R and B will not only be positive, but also positive definite

$$\int_0^l R\eta\eta \mathrm{d}x > \rho_{min} \int_0^l |\eta|^2 \mathrm{d}x, |\eta|^2 = \eta\eta,$$

$$\int_0^l \varepsilon B\varepsilon \mathrm{d}x > b_{min} \int_0^l |\varepsilon|^2 \mathrm{d}x, |\varepsilon|^2 = \varepsilon\varepsilon.$$

This means that from the equations

$$\int_0^l R\eta\eta \mathrm{d}x = 0 \quad \text{and} \quad \int_0^l \varepsilon B\varepsilon \mathrm{d}x = 0$$

it follows that $\eta(x,t) = 0$ and $\varepsilon(x,t) = 0$ for $0 \le x \le l$ at any moment in time $t \ge 0$.

The positive definite operators R and B have positive definite roots $\sqrt{R}$ and $\sqrt{B}$, $\sqrt{R}\sqrt{R} = R$, $\sqrt{B}\sqrt{B} = B$.

Let us introduce some new variables

$$\xi_0 = \sqrt{B}\xi, \quad \eta_0 = \sqrt{R}\eta \tag{1.118}$$

and operators

$$D_0 = (\sqrt{R})^{-1} D (\sqrt{B})^{-1}, \quad D_0^* = (\sqrt{B})^{-1} D^* (\sqrt{R})^{-1},$$

and

$$h_0 = (\sqrt{R})^{-1} h, \quad f_0 = (\sqrt{R})^{-1} f, \quad \varepsilon_0 = (\sqrt{B})^{-1} \varepsilon, \quad e_0 = (\sqrt{B})^{-1} e.$$

Formulas (1.118) bijectively (one-to-one) connect the functions $\xi_0(x,t)$ and $\xi(x,t)$, $\eta_0(x,t)$ and $\eta(x,t)$, respectively. For example, in the problem of the longitudinal vibrations of a rod $\xi = Q$, $\eta = u$, $\sqrt{B} = \sqrt{b}$, $\sqrt{R} = \sqrt{\rho}$, and formula (1.118) can be written as follows

$$Q_0(x,t) = \sqrt{b(x)} Q(x,t), \quad u_0(x,t) = \sqrt{\rho(x)} u(x,t),$$

where $b(x)$ and $\rho(x)$ satisfy the conditions (1.117).

A similar remark regarding bijectivity refers to the operators D and D_0, D^* and D_0^*. Hence, Eqs. (1.19) and (1.20) can be rewritten as follows

$$\begin{aligned} D_0\xi_0 &= h_0, h_0 = -\ddot{\eta}_0 + f_0, \\ D_0^*\eta_0 &= \varepsilon_0, \varepsilon_0 = \xi_0 + e_0, \end{aligned} \tag{1.119}$$

$$\begin{aligned} D_0\xi_0 + \ddot{\eta}_0 - f_0 &= 0, \\ D_0^*\eta_0 - \xi_0 - e_0 &= 0, \end{aligned}$$

while the equation in displacements (1.74) will be

$$\begin{aligned} &N_0\eta_0 + \ddot{\eta}_0 - (f_0 + D_0 e_0) = 0, \\ &N_0 = D_0 D_0^* = (\sqrt{R})^{-1} N (\sqrt{R})^{-1}, \quad 0 \le x \le l. \end{aligned} \tag{1.120}$$

The load $h = h(x, t)$ is called "distributed" if

$$\|h_0\|^2 = \int_0^l |h_0|^2 \mathrm{d}x = \int_0^l hR^{-1}h\mathrm{d}x < \infty, \tag{1.121}$$

where the integral is understood in the sense of Lebesgue, while the function $h_0(x, t)$ is called "square integrable" [3]. Formula (1.121) defines the norm of a function $h_0(x, t)$ and transforms the set of these functions into the closed normed space of such functions. The convenience of introducing the closed normed space of distributed loads consists in the fact that any convergence (in a norm) in this space sequence has a limit belonging to the same space. The latter circumstance can be utilized, for example, in the study of the convergence of approximate methods for solving the equations of the theory of elastic vibrations.

Functions $\xi_0 = \xi_0(x, t)$ will also be considered square integrable

$$\|\xi_0\|^2 = \int_0^l |\xi_0|^2 \mathrm{d}x = \int_0^l \xi B\xi \mathrm{d}x < \infty. \tag{1.122}$$

Moreover, according to (1.119), (1.121)

$$\|h_0\|^2 = \int_0^l |h_0|^2 \mathrm{d}x = \int_0^l |D_0\xi_0|^2 \mathrm{d}x, \quad \|D_0\xi_0\|^2 < \infty.$$

As a result

$$\|\xi_0\|^2 + \|D_0\xi_0\|^2 < \infty. \tag{1.123}$$

Similarly, in the second formula (1.119)

$$\|\eta_0\|^2 < \infty, \tag{1.124}$$

$$\|\varepsilon_0\|^2 = \|D_0^*\eta_0\|^2 < \infty, \tag{1.125}$$

such that

$$\|\eta_0\|^2 + \|D_0^*\eta_0\|^2 < \infty. \tag{1.126}$$

For example, in the problem of transverse vibrations

$$\xi = (Q, M), \quad \xi_0 = (Q_0, M_0),$$
$$Q_0(x,t) = \sqrt{b(x)}Q(x,t), \quad M_0(x,t) = \sqrt{\beta(x)}M(x,t),$$
$$|\xi_0|^2 = \xi_0\xi_0 = \xi B\xi = Q_0Q_0 + M_0M_0 = QbQ + M\beta M,$$

$$\sqrt{B} = \begin{pmatrix} \sqrt{b} & 0 \\ 0 & \sqrt{\beta} \end{pmatrix},$$

and condition (1.122) can be written as follows

$$\int_0^l (QbQ + M\beta M)\mathrm{d}x < \infty.$$

Taking formulas (1.121), (1.122), (1.124), (1.125) and the Bunyakovskii-Schwarz inequality into account, the work of external load

$$\frac{1}{2}(h,\eta) = \frac{1}{2}(h_0,\eta_0) = \frac{1}{2}\int_0^l h_0\eta_0 \mathrm{d}x$$

$$\leq \frac{1}{2}\sqrt{\int_0^l |h_0|^2\mathrm{d}x \int_0^l |\eta_0|^2\mathrm{d}x} = \frac{1}{2}||h_0|| \cdot ||\eta_0|| < \infty,$$

where the brackets denote scalar multiplication of the load by displacement with the integration over the domain (the length of a rod), occupied by an elastic body.

The elastic strain energy

$$\frac{1}{2}(\xi,\varepsilon) \leq \frac{1}{2}||\xi_0|| \cdot ||\varepsilon_0|| < \infty.$$

Note one more condition that load h must obey, along with (1.121), when displacements of a rigid body are possible for the fastening conditions given. These displacements $\widehat{\eta} \neq 0$ satisfy the equation

$$N\widehat{\eta} = 0, \tag{1.127}$$

and, conversely, the nontrivial solutions of Eq. (1.127) describe the rigid displacements to which deformation, equal to zero, correspond

$$\widehat{\varepsilon} = D^* \widehat{\eta} = 0. \tag{1.128}$$

However, by virtue of (1.127)

$$\begin{aligned} \int_0^l B^{-1} \widehat{\varepsilon}\,\widehat{\varepsilon}\,\mathrm{d}x &= \int_0^l B^{-1} D^* \widehat{\eta} D^* \widehat{\eta}\,\mathrm{d}x \\ &= \int_0^l D B^{-1} D^* \widehat{\eta}\widehat{\eta}\,\mathrm{d}x = \int_0^l N \widehat{\eta}\widehat{\eta}\,\mathrm{d}x = 0. \end{aligned} \tag{1.129}$$

When the characteristics of elasticity are bounded in magnitude

$$0 < b_{\min} \le b(x) \le b_{\max} < \infty, \quad 0 < \beta_{\min} \le \beta(x) \le \beta_{\max} < \infty$$

the operator B^{-1} is positive definite. Therefore, (1.128) follows from (1.129). Conversely, if (1.128) is true, then

$$N\widehat{\eta} = DB^{-1}D^*\widehat{\eta} = 0.$$

Now, multiplying scalarly the relation

$$N\eta = h \tag{1.130}$$

by $\widehat{\eta}$ we obtain

$$\int_0^l N\eta\widehat{\eta}\,\mathrm{d}x = \int_0^l h\widehat{\eta}\,\mathrm{d}x.$$

Since the operator N is self-adjoint, then $\int_0^l N\eta\widehat{\eta}\,\mathrm{d}x = \int_0^l \eta N\widehat{\eta}\,\mathrm{d}x = 0$ and the following equality must hold

$$\int_0^l h\widehat{\eta}\,\mathrm{d}x = 0. \tag{1.131}$$

Let $\eta_i(x), \quad i = 1, \ldots, m$, represent the sequence of all possible (for the fastening conditions given) linearly independent movements of a rigid body ($m \le 6$). We will consider them to be R-orthonormal, so that they can be performed using the Schmidt procedure [4]

$$\int_0^l R\eta_i\eta_j \mathrm{d}x = \begin{cases} 1, & i = j, \\ 0, & i \neq j, \end{cases} \quad i,j = 1,\ldots,m. \tag{1.132}$$

Instead of (1.131), we can record the requirement

$$\int_0^l h\eta_i \mathrm{d}x = 0, \quad i = 1,\ldots,m. \tag{1.133}$$

As long as h does not satisfy these conditions (1.133), the load $\tilde{h}$ can be extracted

$$\tilde{h} = h - \frown{h}, \frown{h} = \sum_{i=1}^{m} h_i R\eta_i, \quad h_i = \int_0^l h\eta_i \mathrm{d}x. \tag{1.134}$$

By virtue of (1.132), the condition (1.133)

$$\int_0^l \tilde{h}\eta_i \mathrm{d}x = 0 \tag{1.135}$$

can clearly be performed. That is, the load $\tilde{h}$ does not work on hard (rigid) displacements.

Let us further represent the total displacement in the form $\eta = \tilde{\eta} + \frown{\eta}$, where $\frown{\eta}$ is the hard or rigid displacement and $\tilde{\eta}$ the elastic displacement. Hence, the solution of the equation

$$N\tilde{\eta} = \tilde{h} \tag{1.136}$$

would be unique. If this were not so, there would be two deflections corresponding to one and the same load $\tilde{h}$

$$N\tilde{\eta}_1 = \tilde{h}, \quad N\tilde{\eta}_2 = \tilde{h} \text{ or } N(\tilde{\eta}_1 - \tilde{\eta}_2) = 0.$$

According to (1.127) $\tilde{\eta}_1 - \tilde{\eta}_2 = \frown{\eta}$ and then

$$\int_0^l R(\tilde{\eta}_1 - \tilde{\eta}_2)\frown{\eta} \mathrm{d}x = \int_0^l R\frown{\eta}\frown{\eta} \mathrm{d}x = 0.$$

The operator R is positive definite; therefore, $\widehat{\eta} = \tilde{\eta}_1 - \tilde{\eta}_2 = 0$ and hence $\tilde{\eta}_1 = \tilde{\eta}_2$. The elastic displacement $\tilde{\eta}$ can be represented as

$$\tilde{\eta} = \eta - \widehat{\eta}, \quad \widehat{\eta} = \sum_{i=1}^{m} \beta_i \eta_i, \quad \beta_i = \int_0^l R\eta\eta_i \mathrm{d}x, \tag{1.137}$$

from which it follows that rigid and elastic displacements are R-orthogonal

$$\int_0^l R\widehat{\eta}\tilde{\eta}\mathrm{d}x = 0.$$

Let us return to Eq. (1.74) and represent the inhomogeneous term of the equation in the form

$$\begin{aligned} f &= \tilde{f} + \widehat{f}; \quad \tilde{f} = \sum_{i=m+1}^{\infty} f_i R\eta_i, \\ \widehat{f} &= \sum_{i=1}^{m} f_i R\eta_i, \quad f_i = \int_0^l f\eta_i \mathrm{d}x, \end{aligned} \tag{1.138}$$

and instead of one Eq. (1.74) we write two:

$$N\tilde{\eta} + R(\tilde{\eta})^{\cdot\cdot} - \tilde{f} = 0, \quad N\widehat{\eta} + R(\widehat{\eta})^{\cdot\cdot} - \widehat{f} = 0, \tag{1.139}$$

Moreover, (1.127) can be performed.

If we multiply the second Eq. (1.139) scalarly by the hard displacements $\eta_i(x)$, $i = 1, \ldots, m$, we obtain

$$\ddot{\beta}_i = f_i, \quad i = 1, \ldots, m. \tag{1.140}$$

The first Eq. (1.139) describes the elastic displacements $\tilde{\eta} = \tilde{\eta}(x, t)$, while formulas (1.140) describe the movement of a rigid body

$$\widehat{\eta} = \widehat{\eta}(x, t) = \sum_{i=1}^{m} \beta_i(t)\eta_i(x).$$

Formulas (1.134) and (1.137) are equivalent to the more compact expressions

$$\tilde{h}_0 = h_0 - \sum_{i=1}^{m} h_{0i}\eta_{0i}, \quad h_{0i} = (h_0, \eta_{0i}),$$
$$\tilde{\eta}_0 = \eta_0 - \sum_{i=1}^{m} \beta_{0i}\eta_{0i}, \quad \beta_{0i} = (\eta_0, \eta_{0i}).$$

We have shown that the ratio (1.136) has the unique solution $\tilde{\eta}$ representing elastic displacement, Consequently, there exists the operator A, inverse of N

$$A = N^{-1}, \quad \tilde{\eta} = A\tilde{h}. \tag{1.141}$$

Applying the operator A to the first Eq. (1.139), we obtain the equation

$$\tilde{\eta} + AR(\tilde{\eta})^{\cdot\cdot} - \tilde{\varphi} = 0, \quad \tilde{\varphi} = A\tilde{f}, \tag{1.142}$$

which describes the elastic displacements, whereas relations (1.140) are still valid for the rigid ones.

We have thus shown how equations with a differential operator can lead to the classical form. Analogously, equations with an integral operator can be converted to the classical form as well. In particular, according to (1.114), (1.120) there will be such an integral operator as

$$A_0 = \sqrt{R}A\sqrt{R}, \quad A_0 = (N_0)^{-1} \tag{1.143}$$

1.12 Integral Equations of Harmonic Oscillations for an Unattached Elastic Body

In addition, let us show how to compose the integral equations of periodic oscillations when the fastening conditions of an elastic body permit its movement as a rigid or solid body. For clarity, we do this for the simplest case of periodic longitudinal vibrations of a rectilinear rod with free ends. The equations for the amplitudes of displacement and force in the case of longitudinal oscillations have the form (1.14). The boundary conditions on the ends of a rod are such that

$$Q(0) = Q(l) = 0. \tag{1.144}$$

Let us represent the amplitude of displacement at harmonic oscillations as

$$u(x) = u(0) + \tilde{u}(x), \quad \tilde{u}(0) = 0 \tag{1.145}$$

and introduce

$$\tilde{p}(x) = p(x) + \lambda^2 \rho(x) u(0). \tag{1.146}$$

Then, Eq. (1.14) can be rewritten as follows

$$\begin{aligned} &-\frac{dQ}{dx} - \lambda^2 \rho \tilde{u} - \tilde{p} = 0, \\ &\frac{d\tilde{u}}{dx} - bQ - e = 0, \end{aligned} \tag{1.147}$$

where, according to the first of these equations and conditions (1.144)

$$\int_0^l (\lambda^2 \rho \tilde{u} + \tilde{p}) ds = 0, \tag{1.148}$$

where the coordinate s has exactly the same meaning as the coordinate x.

Let us integrate both Eq. (1.147) over s between the limits 0 and x, whereupon in the first of the obtained equations, using equality (1.148), we replace the integral from 0 to x by the negative integral from x to l. We obtain

$$\begin{aligned} &Q(x) - \int_x^l \big(\lambda^2 \rho(s) \tilde{u}(s) + \tilde{p}(s)\big) ds = 0, \\ &\tilde{u}(x) - \int_0^x (b(s) Q(s) + e(s)) ds = 0 \end{aligned} \tag{1.149}$$

or

$$\begin{aligned} &Q(x) - \int_0^l K(x,s)\big(\lambda^2 \rho(s) \tilde{u}(s) + \tilde{p}(s)\big) ds = 0, \\ &\tilde{u}(x) - \int_0^l K(s,x)(b(s) Q(s) + e(s)) ds = 0, \end{aligned} \tag{1.150}$$

where the introduced kernels are of the form

$$K(x,s) = \begin{cases} 0, & 0 \le s < x, \\ 1, & x \le s \le l, \end{cases} \quad K(s,x) = \begin{cases} 1, & 0 \le s \le x, \\ 0, & x < s \le l. \end{cases}$$

The system of integral equations (1.150) using the notation in (1.145), (1.146) corresponds to the system of differential equations (1.14) under the boundary conditions (1.144) in the problem, where the movements of a rod as a rigid body are permissible.

References

1. Mikhlin, S. G. (1950). *Direct Methods in mathematical physics* (p. 428). Moscow: State Technical Publishing. (in Russian).
2. Timoshenko S. P. (1956) *Strength of materials*, 3rd edn. Van Nostrand, NJ, Part 1, 1955, 442 p.; Part 2, 1956, 572 p.
3. Lusternik, L. A., & Sobolev, V. J. (1968). *Elements of functional analysis* (411 p.). Gordon and Breach.
4. Smirnov V. I. (1964). *Course of higher mathematics* (Vol. IV, 336 p.). Pergamon Press.

Chapter 2
Vibrations of a Three-Dimensional Body, Plate and Ring

2.1 Equations of Three-Dimensional Body Vibrations

Let us use the Cartesian system of rectilinear orthogonal coordinates $x = (x_i), i = 1, 2, 3$. The equations of three-dimensional body vibrations in the curvilinear coordinate system are given in the Appendix.

The stress state at the point of a solid body is characterized by the tensor

$$\underline{\underline{\tau}} = \tau^{ij}, \quad \underline{\underline{\tau}} = \underline{\underline{\tau}}(x, t), \tag{2.1}$$

where $i, j = 1, 2, 3$.

Displacements at any point of the body are determined by the vector

$$\underline{u} = u_i, \quad \underline{u} = \underline{u}(x, t). \tag{2.2}$$

We denote the vector of body forces by $\underline{q} = q^i, \underline{q} = \underline{q}(x, t)$. They consist of given forces $\underline{p} = p^i, \underline{p} = \underline{p}(x, t)$ and inertial forces $(-\rho\ddot{\underline{u}})$, where $\rho = \rho(x)$ is the specific density of a three-dimensional body. Hence, the dynamic equilibrium equations in the vector-tensor form can be written as [2]

$$-\nabla \cdot \underline{\underline{\tau}} - \underline{q} = 0, \quad \underline{q} = \underline{p} - \rho\ddot{\underline{u}}, \tag{2.3}$$

where $\nabla_i = \frac{\partial}{\partial x_i}$.

Deformations through displacements can be expressed using the operation of differentiation

$$\underline{\underline{\varepsilon}} = \frac{1}{2}(\nabla\underline{u} + (\nabla\underline{u})^*), \tag{2.4}$$

where the asterisk denotes the conjugation operation.

V. Fridman, *Theory of Elastic Oscillations*, Foundations of Engineering Mechanics, DOI 10.1007/978-981-10-4786-2_2

In turn, the total strain tensor $\underline{\underline{\varepsilon}}$ can be represented as a sum of the tensors of the elastic strain and the strain from external causes, which are denoted by $\underline{\underline{\phi}}$ and $\underline{\underline{e}}$, respectively. That is

$$\underline{\underline{\varepsilon}} = \underline{\underline{\phi}} + \underline{\underline{e}}. \tag{2.5}$$

The strain $\underline{\underline{e}}$ can be caused, for example, by the piezo effect, electromagnetic action, variable body heating and others.

An elastic body material will be assumed to be isotropic. We partition the stress tensor $\underline{\underline{\tau}}$ into the spherical and deviatoric parts, which are denoted, respectively, by $\underline{\underline{\sigma}}$ and $\underline{\underline{S}}$

$$\underline{\underline{\tau}} = \underline{\underline{\sigma}} + \underline{\underline{S}}. \tag{2.6}$$

The tensor $\underline{\underline{\sigma}}$ is diagonal

$$\underline{\underline{\sigma}} = \begin{pmatrix} \sigma & 0 & 0 \\ 0 & \sigma & 0 \\ 0 & 0 & \sigma \end{pmatrix}, \tag{2.7}$$

where $\sigma = \frac{\tau^{11} + \tau^{22} + \tau^{33}}{3}$ is the mean tensile (compressive) stress.

In which

$$\underline{\underline{S}} = \underline{\underline{\tau}} - \underline{\underline{\sigma}}$$

or in more detail

$$S = \begin{pmatrix} \tau^{11} - \sigma & \tau^{12} & \tau^{13} \\ \tau^{21} & \tau^{22} - \sigma & \tau^{23} \\ \tau^{31} & \tau^{32} & \tau^{33} - \sigma \end{pmatrix}. \tag{2.8}$$

Using formulas (2.7) and (2.8), it is easy to prove that the convolution of the tensors

$$\underline{\underline{\sigma}} \cdot \cdot \underline{\underline{S}} = 0 \tag{2.9}$$

(the convolution of tensors is understood as the sum of the products of their respective components).

We also divide the elastic strain tensor into the spherical and deviatoric parts

$$\underline{\underline{\phi}} = \underline{\underline{\Theta}} + \underline{\underline{\Gamma}}. \tag{2.10}$$

The spherical part of the strain tensor characterizes change of the volume, while the deviatoric part characterizes change of the shape at a given point of an elastic body. According to the law of elasticity

$$\underline{\underline{\Theta}} = k\underline{\underline{\sigma}} \quad \text{and} \quad \underline{\underline{\Gamma}} = \frac{1}{2G}\underline{\underline{S}}, \tag{2.11}$$

where k is the coefficient of volumetric expansion of the material of an elastic body, $k = \frac{1-2\nu}{E}$, E is the tensile (or compressive) modulus of elasticity (Young's modulus); ν is the transverse compression ratio (Poisson's ratio); and G is the shearing modulus of elasticity.

Hence, the total elastic strain will be

$$\underline{\underline{\phi}} = k\underline{\underline{\sigma}} + \frac{1}{2G}\underline{\underline{S}}. \tag{2.12}$$

In view of formulas (2.3)–(2.5) and (2.12) the equations of three-dimensional body vibrations in the vector-tensor form will be recorded as follows

$$\begin{aligned} &-\mathrm{div}\underline{\underline{\tau}} + \rho\ddot{\underline{u}} - \underline{p} = 0, \\ &\mathrm{def}\underline{u} - \left(k\underline{\underline{\sigma}} + \tfrac{1}{2G}\underline{\underline{S}}\right) - \underline{\underline{e}} = 0, \end{aligned} \tag{2.13}$$

where

$$\mathrm{div}\underline{\underline{\tau}} = \nabla \cdot \underline{\underline{\tau}}, \quad \mathrm{def}\underline{u} = \frac{1}{2}(\nabla\underline{u} + (\nabla\underline{u})^*).$$

The objective of this section is to represent the equations of vibrations of a three-dimensional body in the operator form. This can be accomplished by introducing the following designations

$$\begin{aligned} &\xi = \underline{\underline{\tau}}, \quad \eta = \underline{u}, \quad f = p, \quad p = \underline{p}, \quad e = \underline{\underline{e}}, \\ &D\xi = -\mathrm{div}\ \underline{\underline{\tau}}, \quad D^*\eta = \mathrm{def}\underline{u}, \\ &R\ddot{\eta} = \rho\ddot{\underline{u}}, \quad B\xi = k\underline{\underline{\sigma}} + \frac{1}{2G}\underline{\underline{S}}. \end{aligned} \tag{2.14}$$

Hence, Eqs. (2.13) will take the form:

$$\begin{aligned} &D\xi + R\ddot{\eta} - f = 0, \\ &D^*\eta - B\xi - e = 0. \end{aligned} \tag{2.15}$$

The operator equations (2.15) are valid in the whole volume Ω, occupied by the elastic body, excluding its surface Γ.

Let us discover the properties of the operators introduced in the region Ω. At first

$$\begin{gathered}\int\limits_{\Omega} R\eta_1\eta_2 \mathrm{d}\Omega = \int\limits_{\Omega} (\rho\underline{u}_1)\cdot\underline{u}_2 \mathrm{d}\Omega = \int\limits_{\Omega} \underline{u}_1\cdot(\rho\underline{u}_2)\mathrm{d}\Omega = \int\limits_{\Omega} \eta_1 R\eta_2 \mathrm{d}\Omega, \\ \int\limits_{\Omega} R\eta\,\eta \mathrm{d}\Omega = \int\limits_{\Omega} \rho\underline{u}\cdot\underline{u}\mathrm{d}\Omega \geq \rho_{\min}\int\limits_{\Omega} \underline{u}\cdot\underline{u}\mathrm{d}\Omega = \rho_{\min}\int\limits_{\Omega} \eta\,\eta \mathrm{d}\Omega > 0,\end{gathered} \tag{2.16}$$

if $\underline{u} = \eta \neq 0$ in the domain Ω.

According to formulas (2.16) the operator of inertia R is self-adjoint (self-conjugate) and positive definite. Note that the first property follows from the second and does not require a separate proof [4], but is simply taken into account.

Let us now discover the properties of the operator B. In view of formulas (2.6), (2.9) and notations (2.14) we have

$$\begin{aligned}\int\limits_{\Omega} B\xi\,\xi\,\mathrm{d}\Omega &= \int\limits_{\Omega}\left(k\underline{\underline{\sigma}} + \frac{1}{2G}\underline{\underline{S}}\right)\cdot\cdot(\underline{\underline{\sigma}} + \underline{\underline{S}})\mathrm{d}\Omega \\ &= \int\limits_{0}^{l}\left(k\underline{\underline{\sigma}}\cdot\cdot\underline{\underline{\sigma}} + \frac{1}{2G}\underline{\underline{S}}\cdot\cdot\underline{\underline{S}}\right)\mathrm{d}\Omega \geq k_{\min}\int\limits_{\Omega}(\underline{\underline{\sigma}}\cdot\cdot\underline{\underline{\sigma}} + \underline{\underline{S}}\cdot\cdot\underline{\underline{S}})\mathrm{d}\Omega \\ &= k_{\min}\int\limits_{\Omega}\underline{\underline{\tau}}\cdot\cdot\underline{\underline{\tau}}\,\mathrm{d}\Omega = k_{\min}\int\limits_{\Omega}\xi\,\xi\,\mathrm{d}\Omega > 0.\end{aligned} \tag{2.17}$$

Here, the convolution of tensors with themselves is equal to the sum of squares of their elements (i.e. is a positive value). This makes it possible to take the factor $k_{\min}$ out of the integral, since $G = \frac{E}{2(1+\nu)}$ and $k < \frac{1}{2G}$. Formula (2.17) shows that the operator B is positive definite and, hence, self-conjugate (self-adjoint).

Let us proceed to consideration of the properties of the operators D and D^*. We denote the unit vector of the normal to the surface of an elastic body Γ by $\underline{n}$. We also denote distributed forces by $\underline{X} = \underline{n}\cdot\underline{\underline{\tau}}$ and displacement on the surface Γ by $\underline{Y} = \underline{u}$.

Let us write the Clapeyron formula [2]

$$-\int\limits_{\Omega}\mathrm{div}\underline{\underline{\tau}}\cdot\underline{u}\mathrm{d}\Omega + \int\limits_{\Gamma}\underline{X}\cdot\underline{Y}\mathrm{d}\Gamma = \int\limits_{\Omega}\underline{\underline{\tau}}\cdot\cdot\mathrm{def}\underline{u}\mathrm{d}\Omega. \tag{2.18}$$

Using notations (2.14), equality (2.18) can be rewritten as

$$\int_{\Omega} D\xi\,\eta\,\mathrm{d}\Omega = -\int_{\Gamma} XY\mathrm{d}\Gamma + \int_{\Omega} \xi\,D^{*}\eta\,\mathrm{d}\Omega, \tag{2.19}$$

where $X = \underline{X}$ and $Y = \underline{Y}$.

This formula means that the operators D and D^{*} are conjugate in the sense of Lagrange. In the particular case of homogeneous boundary conditions, where there are distributed forces $X = \underline{X} \equiv 0$ on the portion of the surface Γ_1 and displacements $Y = \underline{Y} \equiv 0, \Gamma_1 + \Gamma_2 = \Gamma$, on the surface portion Γ_2, equality (2.19) can be extended into

$$\int_{\Omega} D\xi\,\eta\mathrm{d}\Omega = \int_{\Omega} \xi\,D^{*}\eta\,\mathrm{d}\Omega,$$

that is, the operators D and D^{*} become merely conjugate.

From the system of equations (2.15) it is not hard to proceed to the single equation with respect to displacement η. Really, from the second equations (2.15) it follows that $\xi = B^{-1}(D^{*}\eta - e)$, and substitution of this expression into the first equations (2.15) gives

$$N\eta + R\ddot{\eta} - f = 0, \tag{2.20}$$

where

$$N = DB^{-1}D^{*}\ \text{and}\ f = p + DB^{-1}e.$$

The appearance and properties of their operators show that Eqs. (2.15) and (2.20) are analogous to the equations of oscillations in the simplest case of a straight rod (1.20) and (1.74).

Introducing oscillation equations into operator equations of the single form is particularly advantageous in that similar algorithms can be applied to solve various problems as long as these algorithms can also be written in the operator form.

2.2 Equations of Plate Vibrations

Let us give another example of drawing up a system of operator equations of the type (2.15). Let us consider the transverse vibrations of a thin rectangular plate. We use the Cartesian system of coordinates $x = x_i, i = 1, 2$, which are counted along the perpendicular axes located in the middle surface of a plate. Much more complicated oscillation equations of a shell in a curvilinear coordinate system are considered in the Appendix.

We denote the vector of shearing forces by $\underline{Q} = \underline{Q}(x,t), \underline{Q} = Q^i$ and the tensor by $\underline{\underline{M}} = \underline{\underline{M}}(x,t), \underline{\underline{M}} = M^{ij}$. The tensor's elements at $i = j$ are bending moments, while at $i \neq j$ they are torques (t is time).

Further, let $p = p(x,t)$ be the distributed transverse load on the plate, $m = m(x,t)$ the distributed external moment, $\rho = \rho(x)$ the mass per unit area of the plate and $j = j(x)$ the distributed moment of inertia per unit area of the plate.

The vector-tensor equations of equilibrium of the plate considering inertial forces can be given the form [1, 3]

$$\begin{aligned} -\nabla \cdot \underline{Q} &= q, \quad q = -\rho\ddot{u} + p, \\ \underline{Q} - \nabla \cdot \underline{\underline{M}} &= \underline{\mu}, \quad \underline{\mu} = -j\ddot{\underline{\vartheta}} + \underline{m}, \end{aligned} \tag{2.21}$$

where $u = u(x,t)$ is displacement perpendicular to the middle surface of the plate; $\underline{\vartheta} = \underline{\vartheta}(x,t), \underline{\vartheta} = \vartheta_i$ is the vector of angles of inclination of the middle surface; and $\nabla = \nabla_i$ is the operator of differentiation along the coordinates $x_i, i = 1, 2$.

The connection between displacements and deformations of shearing and bending is as follows

$$\begin{aligned} \nabla u + \underline{\vartheta} &= \underline{\gamma}, \\ \tfrac{1}{2}(\nabla\underline{\vartheta} + (\nabla\underline{\vartheta})^*) &= \underline{\underline{\kappa}}. \end{aligned} \tag{2.22}$$

The total strains of shearing and bending are composed of elastic strains and the strains from external causes

$$\begin{aligned} \underline{\gamma} &= \underline{\varphi} + \underline{g}, \\ \underline{\underline{\kappa}} &= \underline{\underline{\phi}} + \underline{\underline{k}}. \end{aligned} \tag{2.23}$$

The material of the plate will be assumed to be isotropic. We partition the tensor $\underline{\underline{M}}$ into the spherical and deviatoric parts $\underline{\underline{s}}$ and $\underline{\underline{S}}$

$$\underline{\underline{M}} = \underline{\underline{s}} + \underline{\underline{S}}, \tag{2.24}$$

where

$$\underline{\underline{s}} = \begin{pmatrix} s & 0 \\ 0 & s \end{pmatrix}, \quad s = \frac{M^{11} + M^{22}}{2}, \quad \underline{\underline{S}} = \underline{\underline{M}} - \underline{\underline{s}}.$$

The law of elasticity can be written as

$$\underline{\varphi} = b\underline{Q}, \quad \underline{\underline{\phi}} = \alpha\underline{\underline{s}} + \beta\underline{\underline{S}}, \tag{2.25}$$

where [3]

$$b = \frac{12}{5}\frac{(1+\nu)}{Eh}, \quad \alpha = \frac{12(1-\nu)}{Eh^3},$$

$$\beta = \frac{6}{Gh^3}, \quad h = h(x) \text{ is plate thickness}$$

Excluding $q, \underline{\mu}, \underline{\gamma}$ and $\underline{\underline{\kappa}}$ from Eqs. (2.21), (2.22) and using formulas (2.23), (2.25), we obtain a system of equations describing the bending-shearing vibrations of a thin plate in the vector-tensor form

$$\begin{aligned} &-\nabla\cdot\underline{Q}+\rho\ddot{u}-p=0,\\ &\underline{Q}-\nabla\cdot\underline{\underline{M}}+j\ddot{\underline{\vartheta}}-\underline{m}=0, \end{aligned} \tag{2.26}$$

$$\begin{aligned} &\nabla u+\underline{\vartheta}-b\underline{Q}-\underline{g}=0,\\ &\tfrac{1}{2}(\nabla\underline{\vartheta}+(\nabla\underline{\vartheta})^*)-(\alpha\underline{\underline{s}}+\beta\underline{\underline{S}})-\underline{\underline{k}}=0. \end{aligned} \tag{2.27}$$

Equations (2.26) and (2.27) are valid in the region Ω, which coincides with the area of the middle surface of the plate but does not include its boundary Γ. Note that these equations have a certain analogy with the equations of the bending-shearing vibrations of a rod (1.46).

Let us introduce the notation

$$\xi = \begin{pmatrix}\underline{Q}\\ \underline{\underline{M}}\end{pmatrix}, \quad \eta = \begin{pmatrix}u\\ \underline{\vartheta}\end{pmatrix}, \quad f = \begin{pmatrix}p\\ \underline{m}\end{pmatrix}, \quad e = \begin{pmatrix}\underline{g}\\ \underline{\underline{k}}\end{pmatrix}, \tag{2.28}$$

$$D\xi = \begin{pmatrix}-\nabla\cdot\underline{Q}\\ \underline{Q}-\nabla\cdot\underline{\underline{M}}\end{pmatrix}, \quad D^*\eta = \begin{pmatrix}\nabla u+\underline{\vartheta}\\ \frac{1}{2}(\nabla\underline{\vartheta}+(\nabla\underline{\vartheta})^*)\end{pmatrix}, \tag{2.29}$$

$$R\ddot{\eta} = \begin{pmatrix}\rho\ddot{u}\\ j\ddot{\underline{\vartheta}}\end{pmatrix}, \quad B\xi = \begin{pmatrix}b\underline{Q}\\ \alpha\underline{\underline{s}}+\beta\underline{\underline{S}}\end{pmatrix} \tag{2.30}$$

In this notation the tensor-vector equations (2.26), (2.27) take the form of operator equations of the type (2.15).

Let us define the properties of the operators entered. To do this, we draw up expressions for the scalar products of $D\xi$ and η, as well as of ξ and $D^*\eta$

$$\begin{aligned} &\int_\Omega D\xi\eta\, d\Omega = \int_\Omega \left((-\nabla\cdot\underline{Q})u+(\underline{Q}-\nabla\cdot\underline{\underline{M}})\cdot\underline{\vartheta}\right)d\Omega,\\ &\int_\Omega \xi D^*\eta\, d\Omega = \int_\Omega \left(\underline{Q}\cdot(\nabla u+\underline{\vartheta})+\underline{\underline{M}}\cdot\cdot\frac{1}{2}(\nabla\underline{\vartheta}+(\nabla\underline{\vartheta})^*)\right)d\Omega. \end{aligned} \tag{2.31}$$

At the same time, by virtue of the formula for integration by parts

$$\int_{\Omega} (-\nabla \cdot \underline{Q}) u \mathrm{d}\Omega = -\int_{\Gamma} (\underline{n} \cdot \underline{Q}) u \mathrm{d}\Gamma + \int_{\Omega} \underline{Q} \cdot (\nabla u) \mathrm{d}\Omega$$

and the Clapeyron formula

$$\int_{\Omega} (-\nabla \cdot \underline{\underline{M}}) \cdot \underline{\vartheta} \mathrm{d}\Omega = \int_{\Gamma} (-\underline{n} \cdot \underline{\underline{M}}) \cdot \underline{\vartheta} \mathrm{d}\Gamma + \int_{\Omega} \underline{\underline{M}} \cdot\cdot \frac{1}{2} (\nabla \underline{\vartheta} + (\nabla \underline{\vartheta})^*) \mathrm{d}\Omega$$

we can write

$$\int_{\Omega} \Big((-\nabla \cdot \underline{Q}) u + (\underline{Q} - \nabla \cdot \underline{\underline{M}}) \cdot \underline{\vartheta} \Big) \mathrm{d}\Omega = -\int_{\Gamma} \Big((\underline{n} \cdot \underline{Q}) u + (\underline{n} \cdot \underline{\underline{M}}) \cdot \underline{\vartheta} \Big) \mathrm{d}\Gamma + \int_{\Omega} \Big(\underline{Q} \cdot (\nabla u + \underline{\vartheta}) + \underline{\underline{M}} \cdot\cdot \frac{1}{2} (\nabla \underline{\vartheta} + (\nabla \underline{\vartheta})^*) \Big) \mathrm{d}\Omega. \tag{2.32}$$

Then we have

$$\int_{\Gamma} (\underline{n} \cdot \underline{\underline{M}}) \cdot \underline{\vartheta} \mathrm{d}\Gamma = \int_{\Gamma} \left(M\vartheta - H \frac{du}{ds} \right) \mathrm{d}\Gamma = \int_{\Gamma} \left(M\vartheta + \frac{\mathrm{d}H}{\mathrm{d}s} u \right) \mathrm{d}\Gamma, \tag{2.33}$$

when the coordinate s counted along the contour of the plate has been introduced, it can also be accepted that

$$\underline{\vartheta} = \begin{pmatrix} \vartheta \\ -\frac{\mathrm{d}u}{\mathrm{d}s} \end{pmatrix} \tag{2.34}$$

and the bending and torsional moments on the contour of the plate can be denoted by M and H, as well as by ϑ and $\left(-\frac{\mathrm{d}u}{\mathrm{d}s}\right)$, the inclination angles along the normal and tangentially to the contour of the plate, correspondingly. In addition, we have used the formula for integration by parts

$$\int_{\Gamma} H \frac{\mathrm{d}u}{\mathrm{d}s} \mathrm{d}\Gamma = -\int_{\Gamma} \frac{\mathrm{d}H}{\mathrm{d}s} u \mathrm{d}\Gamma,$$

where the summand outside the integral vanishes because the initial and final points of coordinate s coincide.

Further

$$Q = \underline{n} \cdot \underline{Q}$$

which is the shearing force on the plate contour.

Combining this formula and formulas (2.31)–(2.34), we obtain

$$\int_{\Omega} D\xi\,\eta \mathrm{d}\Omega = -\int_{\Gamma} XY \mathrm{d}\Gamma + \int_{\Omega} \xi D^{*}\eta\, \mathrm{d}\Omega,$$

where $X = \underline{X} = \begin{pmatrix} \tilde{Q} \\ M \end{pmatrix}; Y = \underline{Y} = \begin{pmatrix} u \\ \vartheta \end{pmatrix}$; and $\tilde{Q} = Q + \frac{\mathrm{d}H}{\mathrm{d}s}$

That is, the operators D and D^{*} introduced are conjugate in the sense of Lagrange and, if the product on the plate contour $\underline{X} \cdot \underline{Y} = XY = 0$, then these operators are simply conjugate.

Finally

$$\begin{aligned}
\int_{\Omega} B\xi\xi \mathrm{d}\Omega &= \int_{\Omega} \left(b\underline{Q} \cdot \underline{Q} + (\alpha \underline{\underline{s}} + \beta \underline{\underline{S}}) \cdot\cdot (\underline{\underline{s}} + \underline{\underline{S}}) \right) \mathrm{d}\Omega \\
&= \int_{\Omega} (b\underline{Q} \cdot \underline{Q} + \alpha \underline{\underline{s}} \cdot\cdot \underline{\underline{s}} + \beta \underline{\underline{S}} \cdot\cdot \underline{\underline{S}}) \mathrm{d}\Omega \\
&\geq c \int_{\Omega} (\underline{Q} \cdot \underline{Q} + \underline{\underline{s}} \cdot\cdot \underline{\underline{s}} + \underline{\underline{S}} \cdot\cdot \underline{\underline{S}}) \mathrm{d}\Omega = c \int_{\Omega} \xi\xi\, \mathrm{d}\Omega > 0,
\end{aligned}$$

$$\int_{\Omega} R\eta\eta \mathrm{d}\Omega = \int_{\Omega} (\rho\, uu + j\underline{\vartheta})\underline{\vartheta} \mathrm{d}\Omega \geq \mathrm{d} \int_{\Omega} \eta\eta\, \mathrm{d}\Omega > 0,$$

where $c = \min(b, \alpha, \beta)$; and $d = \min(\rho, j)$.

These estimates imply that the operators B and R are positive definite and therefore self-adjoint (self-conjugate).

This is an opportune time to write the equations of the free axially symmetric oscillations of a thin plate, which will be required in what follows. These equations in the polar coordinate system r, ϑ can be recorded on the basis of the formulas in [1] in the form

$$\begin{aligned}
&-\tfrac{1}{r}\tfrac{\partial(rQ_r)}{\partial r} + \rho \ddot{u} = 0, \quad \tfrac{\partial u}{\partial r} + \vartheta = 0, \\
&Q_r - \left(\tfrac{\partial}{\partial r} + \tfrac{1}{r}\right) M_r + \tfrac{1}{r} M_{\vartheta} = 0, \\
&\left(\tfrac{\partial}{\partial r} - \tfrac{1}{r}\right)\vartheta - \beta(M_r - M_{\vartheta}) = 0,
\end{aligned} \tag{2.35}$$

where Q_r, M_r is the shearing force and bending moment on the area element with the normal r; M_{ϑ} is the bending moment on the area element with the normal, which is tangential to the circle with radius r; and u is axial displacement. Moreover,

$\beta = \frac{12(1+\nu)}{Eh^3}$, where $h = h(r)$ is the thickness of the plate. Equations (2.35) can be written in the operator form analogously with what was done for the system of Eq. (1.50). They will be used further in Chap. 10.

2.3 Equations of Ring Vibrations

Let us consider the plane oscillations of a circular ring. We introduce x as the circumferential coordinate, $0 \leq x \leq l, l = 2\pi R$, and R as the radius of the ring. It is considered that the distributed tangential and radial loads $p^1 = p^1(x, t)$ and $p^2 = p^2(x, t)$ are acting on the ring and that the distributed moment $m = m(x, t)$ is acting on the plane of the ring. Suppose further that $\rho = \rho(x)$ and $j = j(x)$ are the distributed inertial characteristics at translational displacements and angular vibrations of the ring. We denote the coefficients of elasticity of a ring at tensile and shearing strain by $b_1 = b_1(x)$ and $b_2 = b_2(x)$, respectively. Furthermore, it is accepted that external causes such as the piezo effect and thermal heating can generate tensile (compression), shearing and bending strains g_1, g_2 and k.

The unknown quantities are $Q^1 = Q^1(x, t)$ and $Q^2 = Q^2(x, t)$ which are tensile and shearing forces; $M = M(x, t)$ the bending moment in the ring; and $u_1 = u_1(x, t), u_2 = u_2(x, t)$ and $\vartheta = \vartheta(x, t)$ the tangential, radial displacements and rotation of the cross-section of the ring subject to vibrations.

The equations of oscillations under tension (compression) of the ring in the tangential direction are

$$\begin{aligned} &-\frac{\partial Q^1}{\partial x} - \frac{Q^2}{R} = q^1, \quad q^1 = -\rho \ddot{u}_1 + p^1, \\ &\frac{\partial u_1}{\partial x} + \frac{u_2}{R} = \varepsilon_1, \quad \varepsilon_1 = b_1 Q^1 + g_1. \end{aligned} \tag{2.36}$$

The equations of dynamic equilibrium and strain equations under plane vibrations of the ring are

$$\begin{aligned} &-\frac{\partial Q^2}{\partial x} + \frac{Q^1}{R} = q^2, \quad q^2 = -\rho \ddot{u}_2 + p^2, \\ &Q^2 - \frac{\partial M}{\partial x} = \mu, \quad \mu = -j\ddot{\vartheta} + m. \end{aligned} \tag{2.37}$$

$$\begin{aligned} &\frac{\partial u_2}{\partial x} - \frac{u_1}{R} + \vartheta = \varepsilon_2, \quad \varepsilon_2 = b_2 Q^2 + g_2, \\ &\frac{\partial \vartheta}{\partial x} = \kappa, \quad \kappa = \beta M + k. \end{aligned} \tag{2.38}$$

Equations (2.36)–(2.38) of ring oscillations differ from the equations of rectilinear axis rod oscillations (1.6), (1.43)–(1.45) by the former being interconnected with each other due to the presence of summands with the factor $\frac{1}{R}$. The easiest way to see the appearance of such summands is to analyze the example of axisymmetric strain of the ring. In this case there arises the ring tensile strain equal to $\varepsilon = \frac{2\pi(R+u)-2\pi R}{2\pi R} = \frac{u}{R}$, where u is radial displacement.

Let us introduce the following designations

$$\begin{gathered} p = \underline{p} = p^i, \quad g = \underline{g} = g_i, \quad Q = \underline{Q} = Q^i, \quad u = \underline{u} = u_i, \quad i = 1, 2, \\ f = (p, m), \quad e = (g, k), \quad \xi = (Q, M), \quad \eta = (u, \vartheta). \end{gathered} \tag{2.39}$$

Let us form the differential operators

$$D = \begin{pmatrix} -\frac{\partial}{\partial x} & \frac{1}{R} & 0 \\ \frac{1}{R} & \frac{\partial}{\partial x} & 0 \\ 0 & 1 & -\frac{\partial}{\partial x} \end{pmatrix} \quad D^* = \begin{pmatrix} -\frac{\partial}{\partial x} & -\frac{1}{R} & 0 \\ \frac{1}{R} & -\frac{\partial}{\partial x} & 0 \\ 0 & 1 & -\frac{\partial}{\partial x} \end{pmatrix}, \tag{2.40}$$

and diagonal algebraic operators of inertia and elasticity

$$R = \mathrm{diag}(\rho, \rho, j) \text{ and } B = \mathrm{diag}(b_1, b_2, \beta). \tag{2.41}$$

Excluding $q^1, q^2, \varepsilon_1, \varepsilon_2, \mu, \kappa$ from Eqs. (2.36)–(2.38) and using the notation introduced for given and unknown quantities (2.39), as well as for the differential and algebraic operators (2.40) and (2.41), we arrive once again at the system of operator equations of the form (1.20). Also confirmed are the properties of the operators contained in these equations

$$\begin{aligned} \int_0^l D\xi\eta \mathrm{d}x &= \int_0^l \left(\begin{pmatrix} -\frac{\partial}{\partial x} & -\frac{1}{R} & 0 \\ \frac{1}{R} & -\frac{\partial}{\partial x} & 0 \\ 0 & 1 & -\frac{\partial}{\partial x} \end{pmatrix} \begin{pmatrix} Q^1 \\ Q^2 \\ M \end{pmatrix} \right) \cdot \begin{pmatrix} u_1 \\ u_2 \\ \vartheta \end{pmatrix} \mathrm{d}x \\ &= \int_0^l \left(\left(-\frac{\partial Q^1}{\partial x} - \frac{Q^2}{R} \right) u_1 + \left(\frac{Q^1}{R} - \frac{\partial Q^2}{\partial x} \right) u_2 + \left(Q^2 - \frac{\partial M}{\partial x} \right) \vartheta \right) \mathrm{d}x \\ &= \int_0^l \left(Q^1 \frac{\partial u_1}{\partial x} - \frac{Q^2}{R} u_1 + \frac{Q^1}{R} u_2 + Q^2 \frac{\partial u_2}{\partial x} + Q^2 \vartheta + M \frac{\partial \vartheta}{\partial x} \right) \mathrm{d}x \\ &= \int_0^l \left(Q^1 \left(\frac{\partial u_1}{\partial x} + \frac{u_2}{R} \right) + Q^2 \left(-\frac{u_1}{R} + \frac{\partial u_2}{\partial x} + \vartheta \right) + M \frac{\partial \vartheta}{\partial x} \right) \mathrm{d}x \\ &= \int_0^l \begin{pmatrix} Q^1 \\ Q^2 \\ M \end{pmatrix} \cdot \left(\begin{pmatrix} \frac{\partial}{\partial x} & \frac{1}{R} & 0 \\ -\frac{1}{R} & \frac{\partial}{\partial x} & 1 \\ 0 & 0 & \frac{\partial}{\partial x} \end{pmatrix} \begin{pmatrix} u_1 \\ u_2 \\ \vartheta \end{pmatrix} \right) \mathrm{d}x = \int_0^l \xi D^* \eta \, \mathrm{d}x. \end{aligned}$$

Now that integration by parts has been performed the terms outside the integral vanish because they have equal values at the lower and upper limits of integration at $x = 0, l$ by virtue of the closedness of the ring.

Further

$$\int_0^l B\xi\xi \mathrm{d}x = \int_0^l \left(\begin{pmatrix} b_1 & 0 & 0 \\ 0 & b_2 & 0 \\ 0 & 0 & \beta \end{pmatrix} \begin{pmatrix} Q^1 \\ Q^2 \\ M \end{pmatrix} \right) \cdot \begin{pmatrix} Q^1 \\ Q^2 \\ M \end{pmatrix} \mathrm{d}x \geq b \int_0^l \xi\xi \mathrm{d}x > 0,$$

analogously

$$\int_0^l R\eta\eta \mathrm{d}x = \int_0^l (\rho(u_1^2 + u_2^2) + j\vartheta^2) \mathrm{d}x \geq c \int_0^l \eta\eta \mathrm{d}x > 0,$$

where $b = \min(b_1, b_2, \beta)$ and $c = \min(\rho, j)$.

Thus, in the problem of vibrations of a ring the adjointness (conjugacy) of the operators D and D^* has been shown, as have the positive definiteness and, therefore, the self-adjointness (self-conjugacy) of the operators B and R.

The equations of elastic body vibrations in the curvilinear coordinate system are given in the Appendix.

References

1. Birger I. A., & Mavlyutov, R. R. (1986). *Strength of materials* (560 pp). Science Publishing House of Physical and Mathematical Literature Fizmatlit, Moscow (in Russian).
2. Lurie A. I. (2005). *Theory of elasticity* (1050 pp.). Berlin: Springer.
3. Timoshenko, S., & Woinowsky-Krieger, S. (1959). *Theory of plates and shells* (635 pp). New York: McGraw-Hill.
4. Mikhlin S. G. (1964). *Variational methods in mathematical physics* (510 pp). Oxford: Pergamon Press.

Chapter 3
Spectral Theory

3.1 Forms and Frequencies of Free Oscillations

By free oscillations we refer to those oscillations that occur in the absence of an external exciting load $f = f(x,t) = 0$ and deformation caused by external causes $e = e(x,t) = 0$, where, as before, x is the coordinate of a point in the region Ω occupied by the elastic body and t is time.

In the operator form the equations of free oscillations can be recorded as follows

$$\begin{aligned} D\xi + R\ddot{\eta} &= 0 \\ D^*\eta - B\xi &= 0, \end{aligned} \tag{3.1}$$

or in displacements

$$N\eta + R\ddot{\eta} = 0, \quad \eta = \eta(x,t) \tag{3.2}$$

Let us assume that an elastic body has finite dimensions and on its border simple homogeneous conditions are given.

Solution of Eqs. (3.1) and (3.2) can be represented as harmonic functions of time

$$\xi(x,t) = \xi(x)\cos(\lambda t + \alpha), \quad \eta(x,t) = \eta(x)\cos(\lambda t + \alpha). \tag{3.3}$$

Hence, for the amplitude of displacements $\eta(x)$ at free oscillations the equations

$$N\eta - \lambda^2 R\eta = 0, \quad \eta = \eta(x) \tag{3.4}$$

can be carried out.

Let us outline the major properties of solutions to this equation, some of which we give without proofs, but explain them with the simplest examples (see, for example, [1, 2]).

V. Fridman, *Theory of Elastic Oscillations*, Foundations of Engineering Mechanics, DOI 10.1007/978-981-10-4786-2_3

Equation (3.4) at certain values of the parameter $\lambda^2 = \lambda_i^2 \geq 0$ has nontrivial solutions $\eta_i = \eta_i(x) \neq 0, \quad i = 1, 2, \ldots$. According to (3.3), (3.4), these solutions determine free harmonic oscillations in that λ_i represent the frequencies and $\eta_i = \eta_i(x), i = 1, 2, \ldots$, the amplitude of free oscillations satisfying the homogeneous equation

$$N\eta_i - \lambda_i^2 R\eta_i = 0. \tag{3.5}$$

The amplitude of free vibrations can be determined to within a multiplier

$$N(c\eta_i) - \lambda_i^2 Rc\eta_i = 0,$$

and, therefore, is also called a form of free oscillations.

From the set of forms that differ by a factor, select one for which

$$\int_\Omega R\eta_i\eta_i \mathrm{d}\Omega = 1, \tag{3.6}$$

and call it R-normalized.

A rod with distributed parameters has an infinite number of degrees of freedom and, accordingly, an infinite number of frequencies and forms of free oscillations. For a rod of finite length the sequence $\lambda_i^2, i = 1, 2, \ldots$, forms a discrete infinite set of nonnegative numbers. The forms of free oscillations representing possible displacements of the body as a hard solid with rigid motions correspond to the frequencies $\lambda_i = 0, \ i = 1, 2, \ldots, m, \ m \leq 6$, whereas the appropriate forms of elastic vibrations correspond to the frequencies $\lambda_i^2 > 0, i = m + 1, \ldots$.

This can be proved by multiplying (3.5) scalarly by $\eta_i, i = 1, 2, \ldots$. We obtain

$$\int_\Omega N\eta_i\eta_i \mathrm{d}\Omega - \lambda_i^2 \int_\Omega R\eta_i\eta_i \mathrm{d}\Omega = 0,$$

such that

$$\lambda_i^2 = \frac{\int_\Omega N\eta_i\eta_i \mathrm{d}\Omega}{\int_\Omega R\eta_i\eta_i \mathrm{d}\Omega}. \tag{3.7}$$

Normalization (3.6) gives

$$\lambda_i^2 = \int_\Omega N\eta_i\eta_i \mathrm{d}\Omega = \int_\Omega DB^{-1}D^*\eta_i\eta_i \mathrm{d}\Omega = \int_\Omega B^{-1}D^*\eta_i D^*\eta_i \mathrm{d}\Omega,$$

such that

$$\lambda_i^2 = \int_{\Omega} B^{-1}\varepsilon_i\varepsilon_i \mathrm{d}\Omega = \begin{cases} 0, & \varepsilon_i(x) = 0, i = 1, 2, \ldots, m, \\ > 0, & \text{at } \varepsilon_i(x) \neq 0, i = m+1, \ldots, \end{cases}$$

where $\varepsilon_i = D^*\eta_i$ is the strain corresponding to the ith form of vibrations η_i.

Let us number the frequencies in nondecreasing order

$$\lambda_1^2 \leq \lambda_2^2 \leq \lambda_3^2 \leq \ldots$$

Equality signs can be used when several linearly independent forms of free oscillations correspond to one and the same magnitude frequency. Different indexes can be assigned to these frequencies. Hence, there can be $\lambda_{i+1}^2 = \cdots = \lambda_{i+n}^2$, but n is necessarily finite. The forms of free oscillations corresponding to different magnitude frequencies are R-orthogonal; that is

$$\int_{\Omega} R\eta_i\eta_j \mathrm{d}\Omega = 0 \quad \text{at } \lambda_i^2 \neq \lambda_j^2. \tag{3.8}$$

This means that equations written for the ith and jth forms can be multiplied, respectively, by η_j and η_i and then integrated over Ω

$$\int_{\Omega} N\eta_i\eta_j \mathrm{d}\Omega - \lambda_i^2 \int_{\Omega} R\eta_i\eta_j \mathrm{d}\Omega = 0,$$
$$\int_{\Omega} N\eta_j\eta_i \mathrm{d}\Omega - \lambda_j^2 \int_{\Omega} R\eta_j\eta_i \mathrm{d}\Omega = 0.$$

Subtracting one equation from the other, by virtue of the self-adjointness of the operators N and R, we obtain

$$(\lambda_i^2 - \lambda_j^2) \int_{\Omega} R\eta_i\eta_j \mathrm{d}\Omega = 0,$$

from which R-orthogonality (3.8) follows at $\lambda_i^2 \neq \lambda_j^2$.

From n linearly independent forms of free oscillations $\eta_{i+1}, \ldots, \eta_{i+n}$, belonging to one and the same frequency, let us move on to the same number of R-orthonormal forms of free oscillations using the algorithm of Schmidt [2]. Let us further assume that the forms of free oscillations, belonging to the equal magnitude frequencies of free oscillations, are subjected to the procedure of R-orthonormality. Then, taking properties (3.6) and (3.8) into account, we can consider that all the forms of free oscillations satisfy the condition of R-orthonormality

$$\int_{\Omega} R\eta_i\eta_j \mathrm{d}\Omega = \delta_{ij}, \quad \delta_{ij} = \begin{cases} 1, & i = j, \\ 0, & i \neq j, \end{cases} \tag{3.9}$$

where δ_{ij} is the Kronecker symbol.

The elastic forms of free oscillations also satisfy an equation with an integral operator

$$\eta_i - \lambda_i^2 AR\eta_i = 0, \quad i = m+1, \ldots, \tag{3.10}$$

which can be received at once by applying the operator A to Eq. (3.5) and taking

$$AN\eta_i = \eta_i.$$

into account at $i = m + 1, \ldots$

Having multiplied scalarly Eq. (3.10) by the function $R\eta_i(x), i = m+1, \ldots,$ we arrive at the following formula

$$\lambda_i^2 = \frac{\int_{\Omega} R\eta_i\eta_i \mathrm{d}\Omega}{\int_{\Omega} R\eta_i AR\eta_i \mathrm{d}\Omega}, \quad i = m+1, \ldots, \tag{3.11}$$

or by carrying out R-normalization (3.6)

$$\lambda_i^2 = \frac{1}{\int_{\Omega} AR\eta_i R\eta_i \mathrm{d}\Omega}, \quad i = m+1, \ldots.$$

As an illustration, let us consider the simplest case of the free longitudinal oscillations of a straight rod. The equation for the displacements of amplitudes of free oscillations $u_i(x)$ is of the form

$$-\frac{\mathrm{d}}{\mathrm{d}x}\left(\frac{1}{b}\frac{\mathrm{d}u_i}{\mathrm{d}x}\right) - \lambda_i^2 \rho u_i = 0, \tag{3.12}$$

and on the ends of the rod at $x = 0, l$

$$\text{either} \quad \frac{1}{b}\frac{\mathrm{d}u_i}{Ex} = 0, \quad \text{or} \quad u_i(x) = 0.$$

For a rod of constant cross-section $(\rho = \rho(x) = \text{const}, b = b(x) = \text{const})$ with the boundary conditions

$$\text{at} \quad x = 0: \quad \frac{1}{b}\frac{\mathrm{d}u_i}{\mathrm{d}x} = 0,$$

$$\text{at} \quad x = l: \quad u_i(x) = 0$$

Equation (3.12) is satisfied with

$$\begin{aligned} \lambda_i^2 &= \frac{\pi^2}{4l^2}\frac{(2i-1)^2}{\rho b}, \\ u_i(x) &= c_i \cos\frac{\pi}{2l}(2i-1)x, \quad i = 1, 2, \ldots, \end{aligned} \tag{3.13}$$

at any c_i. In particular, if $c_i = \sqrt{\frac{2}{\rho l}}$ is chosen, then it will be

$$\int_0^l \rho u_i u_j \mathrm{d}x = \delta_{ij}.$$

According to (3.13)

$$\lambda_{i+1} - \lambda_i = \frac{\pi}{l\sqrt{\rho b}}, \tag{3.14}$$

from which it is evident that the difference between the closest frequencies of a finite length rod is finite and does not depend on the number i.

For a rod with free ends ($Q = 0$ at $x = 0, l$) subject to longitudinal oscillations the frequencies and forms will be

$$\begin{aligned} &\lambda_i^2 = \frac{\pi^2}{l^2}\frac{(i-1)^2}{\rho b}, \quad u_i(x) = A_i \cos\frac{\pi(i-1)}{l}x, \ i = 1, 2, \ldots, \\ &A_i = \sqrt{\frac{2}{\rho l}}, \ i > 1; \quad A_1 = \sqrt{\frac{1}{\rho l}}. \end{aligned} \tag{3.15}$$

The form $u_1(x) = \text{const} = A_1$ (the motion of the rod as a rigid body) belongs to the frequency $\lambda_1 = 0$. Hence, by virtue of (1.3), the deformation of the rod is $\varepsilon_1 = \frac{\mathrm{d}u_1}{\mathrm{d}x} = 0$. The indexes $i = 2, 3, \ldots$ and $\varepsilon_i = \frac{\mathrm{d}u_i}{\mathrm{d}x} \neq 0$ correspond to the elastic forms of free oscillations.

In the examples of the longitudinal oscillations of the rod that we have considered, only one normalized form corresponds to each frequency of free vibrations. Such a situation occurs at vibrations of a certain kind (longitudinal, torsional or transversal) in some of the planes. In the more general case of simultaneous oscillations of different types the coincidence of frequencies corresponding to the various forms of free vibrations cannot be excluded. In other words, several linearly independent forms of free oscillations can belong to one and the same frequency. For a straight rod there can be no more than four such forms of elastic vibrations (by the number of types of vibrations), while there can be no more than six forms of rigid motion, as already mentioned (by the number of possible translational and rotational movements).

3.2 Representation of the Amplitude of Forced Harmonic Vibrations as a Series in the Forms of Free Oscillations

Suppose the frequencies and forms of free oscillations $\lambda_i, \eta_i(x), i = 1, 2, \ldots,$ are known. The forms $\eta_i(x)$ satisfy the condition of R-orthonormality (3.9). We represent the amplitude of forced harmonic oscillations $\eta = \eta(x)$ as the series

$$\eta = \sum_{i=1}^{\infty} \beta_i \eta_i. \tag{3.16}$$

Multiplying the scalar product of both sides of this equality by $R\,\eta_k, k = 1, 2, \ldots,$ we get

$$\int_{\Omega} R\eta\,\eta_k \mathrm{d}\Omega = \int_{\Omega} \left(\sum_{i=1}^{\infty} \beta_i \eta_i \right) R\eta_k \mathrm{d}\Omega = \sum_{i=1}^{\infty} \beta_i \int_{\Omega} R\eta_i \eta_k \mathrm{d}\Omega.$$

Hence, by virtue of (3.9)

$$\beta_i = \int_{\Omega} R\eta\,\eta_i \mathrm{d}\Omega, \quad i = 1, 2, \ldots. \tag{3.17}$$

The amplitude of forced harmonic oscillations under the action of harmonic load with amplitude $f(x)$ and frequency λ satisfies Eq. (1.75); therefore

$$\int_{\Omega} N\eta\eta_i \mathrm{d}\Omega - \lambda^2 \int_{\Omega} R\eta\,\eta_i \mathrm{d}\Omega - \int_{\Omega} f\eta_i \mathrm{d}\Omega = 0, \quad i = 1, 2, \ldots$$

where

$$\int_{\Omega} N\eta\eta_i \mathrm{d}\Omega = \int_{\Omega} \eta N\eta_i \mathrm{d}\Omega = \lambda_i^2 \int_{\Omega} R\eta\,\eta_i \mathrm{d}\Omega \tag{3.18}$$

and in view of formula (3.17)

$$\beta_i = \frac{f_i}{\lambda_i^2 - \lambda^2}, \quad f_i = \int\limits_{\Omega} f \eta_i \mathrm{d}\Omega, \quad i = 1, 2, \ldots. \tag{3.19}$$

Thus, the amplitude of forced harmonic oscillations can be represented as a series in the forms of free oscillations:

$$\eta = \sum_{i=1}^{\infty} \frac{f_i}{\lambda_i^2 - \lambda_2} \eta_i, \tag{3.20}$$

where $\eta = \eta(x)$; $\eta_i = \eta_i(x)$; and f_i, according to formula (3.19), is equal to doubling the workload on the ith form.

The series (3.20) can be rewritten as follows

$$\eta = \sum_{i=1}^{m} \frac{f_i}{-\lambda^2} \eta_i + \sum_{i=m+1}^{\infty} \frac{f_i}{\lambda_i^2 - \lambda^2} \eta_i, \tag{3.21}$$

where the first term represents the amplitude of forced harmonic oscillations of a rigid body ($\lambda_i^2 = 0, i = 1, 2, .., m$), while the second represents the amplitude of elastic vibrations ($\lambda_i^2 > 0, i = m + 1, \ldots$).

Let us note the possible instances:

(1) If $\lambda^2 \neq \lambda_i^2, i = m + 1, m + 2, \ldots$ (i.e. the frequency of forced oscillations does not coincide with any of the frequencies of free oscillations), then Eq. (1.75) has a unique solution, which can be represented as a series (3.21).

(2) $\lambda^2 = \lambda_k^2$ is the frequency of forced oscillations that coincides with one of the natural frequencies of the system (a resonant case). The denominator in formula (3.21), $\lambda_k^2 - \lambda^2 = 0$, means the problem does not have a finite solution when the work of load f on the respective form of free oscillations η_k is not equal to zero

$$f_k = \int\limits_{\Omega} f \eta_k \mathrm{d}\Omega = 2\left(\frac{1}{2}\int\limits_{\Omega} f \eta_k \mathrm{d}\Omega\right) \neq 0.$$

If there is

$$f_k = 0, \tag{3.22}$$

then

$$\eta = \sum_{\substack{i=1 \\ i \neq k}}^{\infty} \frac{f_i}{\lambda_i^2 - \lambda^2} \eta_i + \beta_k \eta_k, \tag{3.23}$$

where $\beta_k = \frac{f_k}{\lambda_k^2 - \lambda^2} = \frac{0}{0}$ is an arbitrary factor.

The above implies the following (the Fredholm alternative). Equation (1.75) either has a unique finite solution at $\lambda^2 \neq \lambda_i^2$, or at $\lambda^2 = \lambda_k^2$ does not have a bounded solution or has an infinite set of solutions (3.23) under the condition (3.22). The latter means that in solution (3.23), along with forced vibrations, free oscillations are also possible, the amplitude of which is determined up to an arbitrary factor β_k.

An analogous approach can also be used to solve the integral Eq. (1.112). In this equation $\eta = \eta(x)$ is the amplitude of forced elastic oscillations. Expanding it to the forms of free oscillations, only the elastic forms $\eta_i = \eta_i(x), i = m+1, \ldots,$ should be kept

$$\eta = \sum_{i=m+1}^{\infty} \beta_i \eta_i. \tag{3.24}$$

Scalarly multiplying Eq. (1.112) by $R\eta_i, i = m+1, \ldots,$ we obtain

$$\int_{\Omega} \eta R \, \eta_i \mathrm{d}\Omega - \lambda^2 \int_{\Omega} AR \, \eta R \, \eta_i \mathrm{d}\Omega - \int_{\Omega} AfR \, \eta_i \mathrm{d}\Omega = 0, \quad i = m+1, \ldots,$$

where

$$\int_{\Omega} AR\eta R\eta_i \mathrm{d}\Omega = \int_{\Omega} R\eta AR\eta_i \mathrm{d}\Omega = \frac{1}{\lambda_i^2} \int_{\Omega} R\eta \, \eta_i \mathrm{d}\Omega,$$

$$\int_{\Omega} AfR \, \eta_i \mathrm{d}\Omega = \int_{\Omega} fAR \, \eta_i \mathrm{d}\Omega = \frac{1}{\lambda_i^2} \int_{\Omega} f\eta_i \mathrm{d}\Omega.$$

Here again we arrive at formula (3.19)

$$\beta_i = \frac{f_i}{\lambda_i^2 - \lambda^2}, \quad \text{but at } i = m+1, \ldots,$$

and for the amplitude of forced elastic vibrations we have

$$\eta = \sum_{i=m+1}^{\infty} \frac{f_i}{\lambda_i^2 - \lambda^2}\eta_i.$$

Hence, there follows a formula for elastic static displacement (1.116) (case $\lambda^2 = 0$)

$$\varphi = \sum_{i=m+1}^{\infty} \frac{f_i}{\lambda_i^2}\eta_i. \tag{3.25}$$

The convergence of the series (3.21) can be significantly improved, if the static displacement $\varphi = Af$ is known. In view of formula (3.25)

$$\begin{aligned}
\eta &= \sum_{i=1}^{\infty} \frac{f_i}{\lambda_i^2 - \lambda^2}\eta_i + \varphi - \varphi \\
&= \varphi + \sum_{i=1}^{\infty} \frac{f_i}{\lambda_i^2 - \lambda^2}\eta_i - \sum_{i=m+1}^{\infty} \frac{f_i}{\lambda_i^2}\eta_i \\
&= \varphi + \sum_{i=1}^{m} \frac{f_i}{-\lambda^2}\eta_i + \sum_{i=m+1}^{\infty} \frac{\lambda^2}{\lambda_i^2} \frac{f_i}{(\lambda_i^2 - \lambda^2)}\eta_i
\end{aligned}$$

or

$$\eta = \varphi + \sum_{i=1}^{m} \frac{f_i}{-\lambda^2}\eta_i + \sum_{i=m+1}^{\infty} \frac{\lambda^2}{\lambda_i^2} \frac{f_i}{(\lambda_i^2 - \lambda^2)}\eta_i. \tag{3.26}$$

At $\lambda_i^2 \to \infty$, members of the series (3.26) decrease proportionately by $\frac{1}{\lambda_i^4}$, while in formula (3.21) they do so proportionately by $\frac{1}{\lambda_i^2}$.

3.3 Bringing Equations to the Classical Form

Note that equations for the forms of free oscillations (3.5) and (3.10) can also be put in the classical form

$$\begin{aligned}
N_0\eta_{0i} - \lambda_i^2\eta_{0i} &= 0, \quad i = 1, 2, \ldots, \\
\eta_{0i} - \lambda_i^2 A_0\eta_{0i} &= 0, \quad i = m + 1, \ldots,
\end{aligned} \tag{3.27}$$

where operators N_0 and A_0 are still as they were in (1.120), (1.143), and the function $\eta_{0i} = \sqrt{R}\eta_i$ has been introduced.

For functions $\eta_{0i}, i = 1, 2, \ldots$, the orthonormality condition (3.9) can be written as

$$\int_{\Omega} \eta_{0i}\eta_{0j}\mathrm{d}\Omega = \delta_{ij}. \tag{3.28}$$

Instead of formulas (3.7), (3.11) we have

$$\begin{aligned} \lambda_i^2 &= \frac{\int_{\Omega} N_0\eta_{0i}\eta_{0i}\mathrm{d}\Omega}{\int_{\Omega}\eta_{0i}\eta_{0i}\mathrm{d}\Omega} = \int_{\Omega} N_0\eta_{0i}\eta_{0i}\mathrm{d}\Omega, \quad i = 1, 2, \ldots, \\ \lambda_i^2 &= \frac{\int_{\Omega}\eta_{0i}\eta_{0i}\mathrm{d}\Omega}{\int_{\Omega} A_0\eta_{0i}\eta_{0i}\mathrm{d}\Omega} = \frac{1}{\int_{\Omega} A_0\eta_{0i}\eta_{0i}\mathrm{d}\Omega}, \quad i = m+1, \ldots. \end{aligned} \tag{3.29}$$

The functions $\eta_{0i} = \eta_{0i}(x), i = 1, 2, \ldots$, are called "eigenfunctions" of the operator N_0, and at $i = m+1, \ldots$ they are also eigenfunctions of the operator A_0. The positive numbers $\lambda_i^2, i = 1, 2, \ldots$, are referred to as eigenvalues or values of the operator N_0, and at $i = m+1, \ldots$ they are also characteristic numbers (values) of the operator A_0. Conversely, $\nu_i^2 = \frac{1}{\lambda_i^2}, i = m+1, \ldots$, are characteristic numbers for the operator N_0 and eigenvalues for the operator A_0. That is, at $i = m+1, \ldots$ the eigenvalues and characteristic numbers (values) of the operators N_0 and A_0 are mutually inverse. The totality of all eigenvalues of the operator is called a "spectrum".

The basis for representation of the amplitude of forced oscillations as a series in the forms of free oscillations can be served by the Hilbert-Schmidt theorem [3]. By virtue of (1.112), (1.143),

$$\eta_0 = A_0 h_0$$

or

$$\eta_0(x) = \int_0^l A_0(x, s)h_0(s)\mathrm{d}s. \tag{3.30}$$

In the more general case

$$\eta_0(x) = \int_{\Omega} A_0(x, s)h_0(s)\mathrm{d}\Omega,$$

where x and s are points belonging to the region Ω

So, for example, in the problem on longitudinal vibrations of a rod $A_0(x, s) = \sqrt{\rho(x)}A(x, s)\sqrt{\rho(s)}$ is a symmetric kernel.

Function $\eta_0(x)$ of the form (3.30) is called "representable" when the core $A_0(x,s)$ is used and, in accordance with the abovementioned Hilbert-Schmidt theorem [2], is expandable in a series in the eigenfunctions of the same kernel

$$\eta_0(x) = \sum_{i=m+1}^{\infty} \beta_i \eta_{0i}(x),$$

$$\eta_{0i}(x) - \lambda_i^2 \int_0^l A_0(x,s)\eta_{0i}(s)\mathrm{d}s = 0.$$

What is more, convergence of the series is uniform.

3.4 Stationary (Periodic) and Nonstationary Elastic Vibrations

Let us return to Eq. (1.74), in which the load $f = f(x,t)$ and the displacement sought $\eta = \eta(x,t)$ are functions of the point coordinates of the elastic body x and time t. Boundary conditions and time conditions should be added to Eq. (1.74). We distinguish two possible cases.

(1) Stationary vibrations. Load is a periodic function of time $f(x,t) = f(x,t+T)$, but a periodic motion is sought. The terms of periodicity are

$$\eta(x,0) = \eta(x,T), \quad \dot{\eta}(x,0) = \dot{\eta}(x,T), \tag{3.31}$$

where T is the period of oscillations

(2) Nonstationary oscillations. At the initial moment $t = 0$ there can be set

$$\eta(x,0) = \varphi_0(x), \quad \dot{\eta}(x,0) = v_0(x). \tag{3.32}$$

The problem of oscillations with the initial conditions given is also called "the initial problem" (the Cauchy problem).

3.5 Oscillations with the Initial Conditions Given

The solution of Eq. (1.74) will be sought as a series in the forms of free oscillations $\eta_i = \eta_i(x), \quad i = 1, 2, \ldots,$ using coefficients of the series $\beta_i = \beta_i(t)$ depending on time $t \geq 0$

$$\eta(x,t) = \sum_{i=1}^{\infty} \beta_i(t)\eta_i(x)$$

or by omitting the arguments

$$\eta = \sum_{i=1}^{\infty} \beta_i \eta_i. \tag{3.33}$$

The coefficients β_i can be determined by multiplying scalarly both sides of Eq. (1.74) by the forms of free oscillations η_i (projection conditions)

$$\int_{\Omega} (N\eta + R\ddot{\eta} - f)\eta_i \mathrm{d}\Omega = 0, \quad i = 1, 2, \ldots. \tag{3.34}$$

Since the forms of free oscillations and the equation concerning the forms and frequencies of free oscillations (3.5) are R-orthonormal

$$\beta_i = \int_{\Omega} R\eta\, \eta_i \mathrm{d}\Omega, \quad \int_{\Omega} R\ddot{\eta}\, \eta_i \mathrm{d}\Omega = \ddot{\beta}_i,$$

$$\int_{\Omega} N\eta\, \eta_i \mathrm{d}\Omega = \int_{\Omega} \eta N\eta_i \mathrm{d}\Omega = \lambda_i^2 \int_{\Omega} R\eta\, \eta_i \mathrm{d}\Omega = \lambda_i^2 \beta_i$$

and thereby

$$\ddot{\beta}_i + \lambda_i^2 \beta_i - f_i = 0, \quad f_i = f_i(t) = \int_{\Omega} f\eta_i \mathrm{d}\Omega, \tag{3.35}$$

$\lambda_i^2 = 0$ at $i = 1, 2, \ldots, m$ and $\lambda_i^2 > 0$ at $i = m + 1, \ldots$

The solution of the inhomogeneous Eq. (3.35) for $i = m + 1, \ldots$ consists of a general solution of the corresponding homogeneous equation

$$\beta_i^{(1)}(t) = c_i \cos \lambda_i t + d_i \sin \lambda_i t$$

and a particular solution of the inhomogeneous equation represented by the integral of Duhamel

$$\beta_i^{(2)}(t) = \frac{1}{\lambda_i} \int_0^t f_i(\tau) \sin \lambda_i (t - \tau) \mathrm{d}\tau,$$

such that

$$\beta_i(t) = c_i \cos \lambda_i t + d_i \sin \lambda_i t + \frac{1}{\lambda_i} \int_0^t f_i(\tau) \sin \lambda_i (t - \tau) \mathrm{d}\tau. \tag{3.36}$$

Clearly, solution $\beta_i^{(2)}(t)$ satisfies the homogeneous initial conditions

$$\beta_i^{(2)}(0) = 0, \quad \dot{\beta}_i^{(2)}(0) = 0. \tag{3.37}$$

For $\lambda_i^2 = 0, i = 1, 2, \ldots, m$

$$\beta_i(t) = c_i + d_i t + \int_0^t f_i(\tau)(t - \tau) \mathrm{d}\tau. \tag{3.38}$$

In the problem with given (specified) initial conditions

$$\beta_i(0) = \int_\Omega R\eta(x, 0)\eta_i(x) \mathrm{d}\Omega, \, \dot{\beta}_i(0) = \int_\Omega R\dot{\eta}(x, 0)\eta_i(x) \mathrm{d}\Omega. \tag{3.39}$$

At the same time, taking (3.37) into account, $c_i = \beta_i(0), d_i = \frac{\dot{\beta}_i(0)}{\lambda_i}$, will be included in formula (3.36) and at $\lambda_i^2 \neq 0, i = m + 1, \ldots,$ we have

$$\beta_i(t) = \beta_i(0) \cos \lambda_i t + \frac{\dot{\beta}_i(0)}{\lambda_i} \sin \lambda_i t + \frac{1}{\lambda_i} \int_0^t f_i(\tau) \sin \lambda_i (t - \tau) \mathrm{d}\tau, \tag{3.40}$$

in the case $\lambda_i^2 = 0, i = 1, 2, \ldots, m$

$$\beta_i(t) = \beta_i(0) + \dot{\beta}_i(0)t + \int_0^t f_i(\tau)(t - \tau) \mathrm{d}\tau. \tag{3.41}$$

Formula (3.41) can also be obtained from formula (3.40) by moving to the limit at $\lambda_i \to 0$ and adopting the disclosure of uncertainty by L'Hôpital

$$\lim_{\lambda_i \to 0} \frac{1}{\lambda_i} \int_0^t f_i(\tau) \sin \lambda_i (t - \tau) \mathrm{d}\tau$$
$$= \lim_{\lambda_i \to 0} \frac{\frac{\mathrm{d}}{\mathrm{d}\lambda_i} \left(\int_0^t f_i(\tau) \sin \lambda_i (t - \tau) \mathrm{d}\tau \right)}{\frac{\mathrm{d}}{\mathrm{d}\lambda_i} (\lambda_i)}$$
$$= \int_0^t f_i(\tau)(t - \tau) \cos \lambda_i (t - \tau) \mathrm{d}\tau |_{\lambda_i = 0} = \int_0^t f_i(\tau)(t - \tau) \mathrm{d}\tau.$$

The limiting expressions for the terms outside the integral at $\lambda_i \to 0$ in formula (3.40) are obvious.

Note there is another method for approximate solution of the problem of non-stationary oscillations. By entering $\alpha_i(t) = \dot{\beta}_i(t)$, we can replace every second-order Eq. (3.35) with two first-order equations

$$\dot{\alpha}_i(t) = -\lambda_i^2 \beta_i(t) + f_i(t),$$
$$\dot{\beta}_i(t) = \alpha_i(t)$$

subject to initial conditions (3.39), where we put $\dot{\beta}_i(0) = \alpha_i(0)$. This initial problem can be solved by known numerical methods, such as the Runge-Kutta method.

3.6 Periodic Oscillations

On the basis of (3.31)

$$\beta_i(0) = \beta_i(T), \quad \dot{\beta}_i(0) = \dot{\beta}_i(T). \tag{3.42}$$

Substituting expressions (3.36) for β_i in the terms of periodicity (3.42), we obtain a system of two algebraic equations for the coefficients $c_i, d_i, i = m + 1, \ldots$

$$c_i = c_i \cos \lambda_i T + d_i \sin \lambda_i T + \frac{1}{\lambda_i} \int_0^T f_i(\tau) \sin \lambda_i (T - \tau) \mathrm{d}\tau,$$

$$\lambda_i d_i = -\lambda_i c_i \sin \lambda_i T + \lambda_i d_i \cos \lambda_i T + \int_0^T f_i(\tau) \cos \lambda_i (T - \tau) \mathrm{d}\tau.$$

Hence

$$c_i = \frac{1}{2\lambda_i \sin\frac{\lambda_i T}{2}} \int_0^T f_i(\tau) \cos \lambda_i \left(\tau - \frac{T}{2}\right) \mathrm{d}\tau,$$
$$d_i = \frac{1}{2\lambda_i \sin\frac{\lambda_i T}{2}} \int_0^T f_i(\tau) \sin \lambda_i \left(\tau - \frac{T}{2}\right) \mathrm{d}\tau. \tag{3.43}$$

$\beta_i^{\&} = f_i(t)$ at $\lambda_i = 0$ and the solution of Eq. (3.35) will be

$$\beta_i(t) = c_i + d_i t + \int_0^t f_i(\tau)(t - \tau)\mathrm{d}\tau,$$

$$\dot{\beta}_i(t) = d_i + \int_0^t f_i(\tau)\mathrm{d}\tau.$$

From the periodicity conditions (3.42), it follows that

$$c_i = c_i + d_i T + \int_0^T f_i(\tau)(T - \tau)\mathrm{d}\tau, \quad d_i = d_i + \int_0^T f_i(\tau)\mathrm{d}\tau,$$

and hence

$$d_i T + \int_0^T f_i(\tau)(T - \tau)\mathrm{d}\tau = 0,$$

$$\int_0^T f_i(\tau)\mathrm{d}\tau = 0.$$

It can be seen from (3.43) that at $\lambda_i T = 2\pi k$, (k an integer), the problem does not have a finite solution. To avoid such a result, it is necessary to take into account the action of forces of friction.

3.7 Oscillations of a Rod Under the Action of Concentrated Force

For concreteness, let us suppose that the rod is thin and performs bending vibrations under the action of the lateral force $P(t)$, depending on time, applied in the cross-section of the rod with coordinate s.

Let us consider some typical problems on periodic and nonstationary vibrations.

(1) The periodic oscillation problem.

Let the force $P(t)$ be a harmonic function of time

$$P(t) = P\cos(\lambda t + \alpha), \quad P = \text{const.}$$

The equivalent distributed lateral load is $p(x,t) = P(t)\delta(x-s)$, or $p(x,t) = p(x)\cos(\lambda t+\alpha)$, where $p(x) = P\delta(x-s), s = \text{const}$; and $\delta(x-s)$ is the Dirac delta function (see Chap. 6).

Solution of the harmonic oscillation problem can be determined by formulas (3.16), (3.19)–(3.21)

$$\eta(x,t) = \eta(x)\cos(\lambda t + \alpha),$$

$$\eta(x) = \sum_{i=1}^{\infty} \beta_i \eta_i(x), \quad \beta_i = \frac{f_i}{\lambda_i^2 - \lambda^2}, \quad f_i = \int_0^l f\eta_i \mathrm{d}x,$$

hence

$$f(x) = p(x) = P\delta(x-s), \tag{3.44}$$

$$f_i = f_i(s) = P\int_0^l \delta(x-s)\eta_i(x)\mathrm{d}x = P\eta_i(s),$$

such that

$$\eta(x) = P\sum_{i=1}^{\infty} \frac{\eta_i(x)\eta_i(s)}{\lambda_i^2 - \lambda^2}. \tag{3.45}$$

We introduce

$$\Gamma(x,s,\lambda^2) = \sum_{i=1}^{\infty} \frac{\eta_i(x)\eta_i(s)}{\lambda_i^2 - \lambda^2}. \tag{3.46}$$

This core is called the “resolvent” and represents the amplitude of harmonic oscillations in the cross-section of a rod with coordinate x, caused by the action of harmonic concentrated force with unit amplitude applied in the cross-section with coordinate s, at the frequency of forced oscillations λ. Then formula (3.45) can be written as follows

$$\eta(x) = P\Gamma(x, s, \lambda^2). \tag{3.47}$$

If an elastic body is fastened such that the motions of an absolutely rigid body are impossible, then in the static case when $\lambda = 0$

$$\Gamma(x, s, 0) = \sum_{i=1}^{\infty} \frac{\eta_i(x)\eta_i(s)}{\lambda_i^2}$$

and the resolvent takes the value of a purely static elastic deflection (see Sect. 3.2)

$$A(x, s) = \Gamma(x, s, 0) = \sum_{i=1}^{\infty} \frac{\eta_i(x)\eta_i(s)}{\lambda_i^2}. \tag{3.48}$$

This function is also called “the influence function” (of the static unit force applied in the cross-section s on the displacement in the cross-section x).

At the same time, the resolvent $\Gamma(x, s, \lambda^2)$ of formula (3.46) can be used to compute the amplitude of harmonic oscillations $\eta(x)$ under the action of distributed harmonic load $p(s), 0 \leq s \leq l$, acting with frequency λ

$$\eta(x) = \eta(x, \lambda^2) = \int_0^l \Gamma(x, s, \lambda^2) p(s) \mathrm{d}s. \tag{3.49}$$

Hence, the total amplitude of vibrations can comprise both elastic displacements and oscillations as rigid body motions.

(2) The initial problem.

When a beam in a state of rest at the initial time $t = 0$ is subjected to the sudden loading of concentrated force, which further keeps a constant value

$$P(t) = \begin{cases} 0, & t < 0, \\ P = \text{const}, & t \geq 0. \end{cases}$$

There is an equivalent load on the beam at $t \geq 0$

$$f(x) = p(x) = P\delta(x - s).$$

Deflection of the beam $\eta(x,t)$ is sought in the form of a series (3.33) with coefficients $\beta_i(t)$ determined by formulas (3.36).

The initial conditions at $t = 0$ are

$$\eta(x,0) = 0, \quad \dot{\eta}(x,0) = 0, \tag{3.50}$$

and, therefore, according to (3.39)

$$\beta_i(0) = 0, \quad \dot{\beta}_i(0) = 0. \tag{3.51}$$

Then, by virtue of (3.51) at $t > 0$ we have

$$\beta_i(t) = \frac{1}{\lambda_i}\int_0^t f_i(\tau)\sin\lambda_i(t-\tau)\mathrm{d}\tau, \tag{3.52}$$

where $f_i(t) = f_i = \int_0^l p(x)\eta_i(x)\mathrm{d}x = P\int_0^l \delta(x-s)\eta_i(x)\mathrm{d}x = P\eta_i(s)$ does not depend on t.

That is, $\beta_i(t) = \frac{P\eta_i(s)}{\lambda_i}\int_0^t \sin\lambda_i(t-\tau)\mathrm{d}\tau = P\frac{\eta_i(s)}{\lambda_i^2}(1-\cos\lambda_i t)$.

Finally

$$\eta(x,t) = P\sum_{i=1}^{\infty}\frac{\eta_i(x)\eta_i(s)}{\lambda_i^2}(1-\cos\lambda_i t). \tag{3.53}$$

Recall that static deflection under the action of concentrated force P is equal to

$$\varphi(x) = PA(x,s) = P\sum_{i=1}^{\infty}\frac{\eta_i(x)\eta_i(s)}{\lambda_i^2}. \tag{3.54}$$

Comparing (3.53) and (3.54), we can see that the members of the series (3.53) double in value at $t = \frac{\pi k}{\lambda_i}, k = 2n-1, n = 1,2,\ldots$.

(3) The problem of force motion along the beam.

The location coordinate of force $P = P(t) = \text{const}$ at an instant of time $t > 0$ is equal to $s = vt$, where $v = \text{const}$ is the speed at which force moves along the beam. The equivalent distributed lateral load on the beam is

$$p(x,s) = P\delta(x-s) \quad \text{or} \quad p(x,t) = P\delta(x-vt).$$

At the initial time $t = 0$ let the beam be at rest; that is, let conditions (3.50) and (3.51) be satisfied. The coefficients β_i can be determined with formula (3.52) when

$$f_i(t) = \int_0^l p(x,t)\eta_i(x)\mathrm{d}x = \int_0^l P\delta(x - vt)\eta_i(x)\mathrm{d}x = P\eta_i(vt).$$

Thus,

$$\beta_i(t) = \frac{P}{\lambda_i}\int_0^t \eta_i(v\tau)\sin\lambda_i(t-\tau)\mathrm{d}\tau$$

and

$$\eta(x,t) = P\sum_{i=1}^{\infty}\frac{\eta_i(x)}{\lambda_i}\int_0^t \eta_i(v\,\tau)\sin\lambda_i(t-\tau)\mathrm{d}\tau. \tag{3.55}$$

In the more general case, when the magnitude of the concentrated force applied and the speed of its movement along the beam are not constant

$$\eta(x,t) = \sum_{i=1}^{\infty}\frac{\eta_i(x)}{\lambda_i}\int_0^t P(\tau)\eta_i(s(\tau))\sin\lambda_i(t-\tau)\mathrm{d}\tau, \quad s(0) = 0,\ \ 0 \le s(t) \le l. \tag{3.56}$$

3.8 Iterative Method for Determination of the First Form and Frequency of Free Elastic Oscillations

For the sake of ease, we prove the convergence of the iterative process of determining the forms and frequencies of free oscillations for the simplest case, when the kernels of integral equations are bounded, as, for example, in Sects. 1.8 and 1.9 (longitudinal and transverse vibrations of a rod).

Let us refer again to Eqs. (3.10) and to formulas (3.11), which we write as

$$\eta_i = Ap_i, \quad p_i = \lambda_i^2 R\eta_i, \quad \lambda_i^2 = \frac{(R\eta_i, \eta_i)}{(AR\eta_i, R\eta_i)}, \tag{3.57}$$

where scalar products have been introduced

$$(R\eta_i, \eta_i) = \int_\Omega R\eta_i\eta_i\mathrm{d}\Omega, \quad (AR\eta_i, R\eta_i) = \int_\Omega AR\eta_i R\eta_i\mathrm{d}\Omega.$$

Equation (3.10) at $\lambda_i^2 = 0$ has only the trivial solution $\eta_i = \eta_i(x) \equiv 0$. That is, nontrivial solutions of Eq. (3.10) are the forms of elastic vibrations, for which $\lambda_i^2 > 0$.

Let us enumerate the forms of free elastic vibrations

$$\eta_i = \eta_i(x), \quad i = 1, 2, 3\ldots, \tag{3.58}$$

respectively, to the frequencies of free elastic oscillations

$$0 < \lambda_1^2 \le \lambda_2^2 \le \lambda_3^2 \le \ldots \tag{3.59}$$

Let us arrange an iterative process that can identify the forms of free oscillations, one that corresponds to the first (the lowest in absolute value) frequency of free oscillations.

Construction of an iterative process is important, and with this in mind formulas (3.57) show that a form of free oscillations can be regarded as the static deflection η_i under the action of inertial load p_i.

Let us first assume that in the sequence (3.59) the first r frequencies are equal among themselves; that is, to one and the same frequency of free oscillations there correspond r linearly independent forms of free oscillations. We denote these forms as $\eta_i^*, i = 1, 2, \ldots, r$.

Let us choose inertial load $p^{(0)} = p^{(0)}(x)$ as the initial (the zero) approximation and the corresponding static displacement

$$\eta^{(0)} = Ap^{(0)}, \tag{3.60}$$

where the superscript indicates the number of approximations. Suppose further that displacement $\eta^{(0)}$ contains the linear combination of the forms of free oscillations $\eta_i^*, i = 1, 2, \ldots, r$, equal to $\sum_{i=1}^{r} \alpha_i \eta_i^*$. However, any linear combination of natural forms of vibrations, belonging to one and the same natural frequency, is also a form of free oscillations with the same frequency. Therefore, in the initial approximation $\eta^{(0)}$ only one normalized form of free oscillations $\eta_1 = \eta_1(x)$ is entered

$$a_1^{(0)} \eta_1 = \sum_{i=1}^{r} \alpha_i \eta_i^*, \quad (p^{(0)}, \eta_1) \neq 0, \tag{3.61}$$

and prove that successive approximations of the iterative process considered below have just this form in the limit.

Let us suppose that there belongs only one form of free oscillations to the first frequency of free oscillations. Further, we will not consider the first frequency of free oscillations as the multiple and renumber the natural frequencies of vibrations so that instead of the sequence (3.59) we will have

$$0<\lambda_1^2<\lambda_2^2\leq\lambda_3^2\leq\ldots \tag{3.62}$$

In accordance with this numbering of natural frequencies, the numbering of natural forms will also be established.

By virtue of formulas (3.60) and (3.61) there will be also

$$\begin{aligned} a_1^{(0)} &= (R\eta^{(0)},\eta_1) = (RAp^{(0)},\eta_1) \\ &= (p^{(0)},AR\eta_1) = \frac{1}{\lambda_1^2}(p^{(0)},\eta_1) \neq 0. \end{aligned} \tag{3.63}$$

Formula (3.60) can also be written as (see formula 1.113):

$$\eta^{(0)}(x) = \int_0^l A(x,s)p^{(0)}(s)\mathrm{d}s.$$

That is, $\eta^{(0)}(x)$ is a function representable through the core $A(x,s)$. Such a function can be expanded into a series of the forms of free oscillations (eigenfunctions of this kernel)

$$\eta^{(0)}(x) = \sum_{i=1}^{\infty} a_i^{(0)}\eta_i(x). \tag{3.64}$$

Kernel $A(x,s)$ is presumed to be symmetrical and continuous. Hence, the series (3.64) converges regularly; that is, the series consisting of the absolute values of its members converges uniformly [2].

In the first approximation the inertial load is equal to

$$(\lambda^2)^{(0)}R\,\eta^{(0)}, \tag{3.65}$$

where the frequency of free oscillations

$$(\lambda^2)^{(0)} = \frac{(R\,\eta^{(0)},\eta^{(0)})}{(AR\,\eta^{(0)},R\,\eta^{(0)})}.$$

Successive approximations for the forms of free oscillations will be calculated as the static displacement under the action of inertial load.

However, since the forms of free oscillations are determined up to scale and would still be normalized, then the scale of the inertial load can also be chosen from the condition for convenience of calculation and ease of formula. Therefore, instead of formula (3.65), we will use a more simple formula

$$p^{(1)} = R\eta^{(0)}$$

for the inertial load. We define the static displacement under the action of this load as that used in the first approximation of the iterative process.

According to (3.64), $\eta^{(0)} = \sum_{i=1}^{\infty} a_i^{(0)} \eta_i$ and, by virtue of the equations for the forms of free oscillations (3.10), $AR\,\eta_i = \frac{\eta_i}{\lambda_i^2}$. Therefore, a series expansion of the function $\eta^{(1)} = \eta^{(1)}(x)$ in the forms of free oscillations will be

$$\eta^{(1)} = \sum_{i=1}^{\infty} a_i^{(1)} \eta_i, \quad \text{where} \quad a_i^{(1)} = \frac{a_i^{(0)}}{\lambda_i^2}. \tag{3.66}$$

Analogously, the inertial load $p^{(2)} = R\,\eta^{(1)}$, frequency $(\lambda^2)^{(1)} = \frac{(R\,\eta^{(1)}, \eta^{(1)})}{(AR\,\eta^{(1)}, R\,\eta^{(1)})}$ and displacement $\eta^{(2)} = Ap^{(2)}$ are calculated.

For the function $\eta^{(2)} = \eta^{(2)}(x)$, series expansion in the forms of free oscillations will be written as

$$\eta^{(2)} = \sum_{i=1}^{\infty} a_i^{(2)} \eta_i, \quad \text{where} \quad a_i^{(2)} = \frac{a_i^{(1)}}{\lambda_i^2} = \frac{1}{\lambda_i^2}\left(\frac{a_i^{(0)}}{\lambda_i^2}\right) = \frac{a_i^{(0)}}{(\lambda_i^2)^2}.$$

Similarly, we can also calculate the following successive approximations with the formulas

$$\begin{aligned} p^{(n)} &= R\eta^{(n-1)}, \quad (\lambda^2)^{(n-1)} = \frac{(R\,\eta^{(n-1)}, \eta^{(n-1)})}{(AR\,\eta^{(n-1)}, R\,\eta^{(n-1)})}, \\ \eta^{(n)} &= Ap^{(n)}. \end{aligned} \tag{3.67}$$

Hence

$$\eta^{(n)} = \sum_{i=1}^{\infty} a_i^{(n)} \eta_i, \quad \text{where} \quad a_i^{(n)} = \frac{a_i^{(0)}}{(\lambda_i^2)^n}. \tag{3.68}$$

By allocating just the first form of free oscillations, which corresponds to the first (lowest) frequency of free oscillations λ_1, the expansion (3.68) can be represented as

$$\eta^{(n)} = \sum_{i=1}^{\infty} a_i^{(n)} \eta_i = \sum_{i=1}^{\infty} \frac{a_i^{(0)}}{(\lambda_i^2)^n} \eta_i = \frac{a_1^{(0)}}{(\lambda_1^2)^n}\left(\eta_1 + \sum_{i=2}^{\infty} \frac{(\lambda_1^2)^n a_i^{(0)}}{(\lambda_i^2)^n a_1^{(0)}} \eta_i\right). \tag{3.69}$$

Moreover, by virtue of orthonormality of the forms of free oscillations

$$\sqrt{(R\eta^{(n)},\eta^{(n)})} = \sqrt{\sum_{i=1}^{\infty}(a_i^{(n)})^2} = \frac{\left|a_1^{(0)}\right|}{(\lambda_1^2)^n}\sqrt{1+\sum_{i=2}^{\infty}\left(\frac{(\lambda_1^2)^n a_i^{(0)}}{(\lambda_i^2)^n a_1^{(0)}}\right)^2}. \tag{3.70}$$

However, because of the uniform convergence of the series $\sum_{i=2}^{\infty}\left|a_i^0\eta_i\right|$ and formulas (3.62), (3.63), in formula (3.69) there will be

$$\left|\sum_{i=2}^{\infty}\frac{(\lambda_1^2)^n a_i^{(0)}}{(\lambda_i^2)^n a_1^{(0)}}\eta_i\right| \le \frac{(\lambda_1^2)^n}{(\lambda_2^2)^n\left|a_1^{(0)}\right|}\sum_{i=2}^{\infty}\left|a_i^{(0)}\eta_i\right| \to 0, \quad n\to\infty. \tag{3.71}$$

Moreover, in formula (3.70)

$$\begin{aligned}\sum_{i=2}^{\infty}\left(\frac{(\lambda_1^2)^n a_i^{(0)}}{(\lambda_i^2)^n a_1^{(0)}}\right)^2 &= \frac{(\lambda_1^2)^n}{(\lambda_2^2)^n}\frac{1}{(a_1^{(0)})^2}\sum_{i=2}^{\infty}\frac{(\lambda_2^2)^n}{(\lambda_i^2)^n}(a_i^{(0)})^2\\ &\le \frac{(\lambda_1^2)^n}{(\lambda_2^2)^n}\frac{1}{(a_1^{(0)})^2}\sum_{i=2}^{\infty}(a_i^{(0)})^2 \to 0, \quad n\to\infty,\end{aligned} \tag{3.72}$$

since $\frac{(\lambda_2^2)^n}{(\lambda_i^2)^n} \le 1$ at $i = 2,3,\ldots$.

Thus, according to formulas (3.69)–(3.72), uniform convergence takes place

$$\frac{\eta^{(n)}}{\sqrt{(R\eta^{(n)},\eta^{(n)})}} \to \eta_1 \quad \text{at } n\to\infty. \tag{3.73}$$

Then, there will also be

$$(\lambda^2)^{(n)} = \frac{(R\,\eta^{(n)},\eta^{(n)})}{(AR\,\eta^{(n)},R\,\eta^{(n)})} \to \frac{(R\,\eta_1,\eta_1)}{(AR\,\eta_1,R\,\eta_1)} \to \lambda_1^2 \quad \text{at } n\to\infty. \tag{3.74}$$

3.9 Determination of Higher Forms and Frequencies of Free Oscillations

Let us start by determining the second form of free oscillations. We presume that the first form has already been found. Having repeated the previous arguments, we consider that to the second frequency of free oscillations there corresponds only one form of free oscillations $\eta_2 = \eta_2(x)$. Let us take $p^{(0)} = p^{(0)}(x)$ as the initial inertial load such that

$$(p^{(0)}, \eta_2) \neq 0, \tag{3.75}$$

and "clean up" $p^{(0)}$ from the load, corresponding to the first natural form of vibrations

$$p_2^{(0)} = p^{(0)} - (p^{(0)}, \eta_1)R\eta_1.$$

Taking the scalar product of both sides of this equation and the first natural form of the oscillations η_1 and considering R-normalization of this form $(R\eta_1, \eta_1) = 1$, we can verify that $(p_2^{(0)}, \eta_1) = 0$.

Let us repeat the iteration procedure (3.67) described above. This means the series (3.64), (3.66) and (3.68) would begin with index 2, not index 1. As a result, we arrive at the assertion that

$$\frac{\eta^{(n)}}{\sqrt{(R\eta^{(n)}, \eta^{(n)})}} \to \eta_2$$

and correspondingly

$$\lambda^2(\eta^{(n)}) \to \lambda_2^2 \quad \text{at } n \to \infty. \tag{3.76}$$

Next, choose the initial load $p^{(0)} = p^{(0)}(x)$ such that

$$(p^{(0)}, \eta_3) \neq 0, \tag{3.77}$$

and "purify" it of inertial loads that correspond to the first and second forms of free oscillations

$$p_3^{(0)} = p^{(0)} - (p^{(0)}, \eta_1)R\eta_1 - (p^{(0)}, \eta_2)R\eta_2.$$

Repeating the iterative procedure again, we obtain

$$\frac{\eta^{(n)}}{\sqrt{(R\eta^{(n)}, \eta^{(n)})}} \to \eta_3 \quad \text{and} \quad \lambda^2(\eta^{(n)}) \to \lambda_3^2 \tag{3.78}$$

and so on for the higher forms and frequencies of free oscillations.

As already mentioned, the iterative method outlined allows determining only those forms of free oscillations that are contained in the initial zero-order approximation. In other words, if it is required to determine the first n free vibration frequencies and at least one appropriate form of free oscillations, then the single zero-order approximation should be chosen in such a way that conditions of the form (3.61), (3.75), (3.77) can be carried out at once for all n forms of free oscillations

$$(p^{(0)}, \eta_i) \neq 0, \quad i = 1, \ldots, n.$$

In conclusion, we note that by calculating successive approximations of the forms of free oscillations it is possible to use such a formula

$$N\eta^{(n)} = p^{(n)} \tag{3.79}$$

instead of the formula $\eta^{(n)} = Ap^{(n)}$.

Hence, there is no need to construct an integral operator, which is difficult to implement in complicated problems. At the same time, there are many methods for solving inhomogeneous differential equations such as Eq. (3.79).

Note also that when an elastic body is not fastened completely successive approximations by formula (3.79) at each step should be "cleansed" of possible rigid motions, which are the forms of free oscillations corresponding to the zero frequency.

References

1. Lusternik, L. A., & Sobolev, V. J. (1968). *Elements of functional analysis* (411 pp). Gordon and Breach.
2. Smirnov, V. I. (1964). *Course of higher mathematics* (Vol. IV, 336 pp). Oxford: Pergamon Press.

Chapter 4
Variational and Projection Methods for Solving Vibration Theory Equations

Exact solutions of equations of the theory of vibrations can only be constructed for a limited class of problems under homogeneous properties of an elastic body. However, if the elastic, inertial and dissipative properties are variable in coordinate, then there is a need to use approximate methods to solve equations of the theory of vibrations.

The important class of approximate methods for solving functional equations are those based on the use of variational principles. Such techniques are also referred to as variational methods. A variational principle states the interrelation of equations of the problem with the condition of stationarity of the corresponding functional. The equations are replaced by getting rid of the first variation of the functional. Hence, if the solution of the equation provides the functional with not only a stationary but also an extreme value, then there occurs a fixed sign definiteness of the second variation of the functional. In the case of the functional minimum the second variation of the functional is positive, while at the functional maximum it is negative.

The variational method comprises not only the condition of stationary (extremum) of the problem functional. It also uses the representation of the approximate solution and its variation in the form of an expansion in finite series in so-called "coordinate functions". Coordinate functions have to be linearly independent and possess certain properties of the solution sought; namely, those that belong to the domain of definition of the operators of equations and that satisfy all the boundary conditions or any part thereof. Such an approach was first proposed by W. Ritz in [8].

Construction of an algorithm of the variational method for approximate solution of the original problem can be carried out in two steps. First, in the variational condition an approximate representation is substituted in the form of a series only for variation of the solution. The property of arbitrariness of variation has such a consequence as the arbitrariness of a finite number of coefficients of the series. Therefore, the single condition of stationarity of the functional of the problem can be approximately replaced with a finite number of so-called "projection

V. Fridman, *Theory of Elastic Oscillations*, Foundations of Engineering Mechanics, DOI 10.1007/978-981-10-4786-2_4

conditions". Each projection condition represents nothing more than vanishing (equality to zero) of the scalar product of the residual of the equation and of the corresponding coordinate function. The method based on the use of projection conditions, bypassing the condition of stationarity of the functional of the problem, is called the "projection method". The concept was first proposed by Galerkin [4]. Possibilities for projection methods are considerably wider; they can be used even when it is impossible to construct the functional whose stationarity condition would be equivalent to the original equation.

For series expansion and variation of the solution, we propose to use the same coordinate functions in the Galerkin method. G.I. Petrov expanded the capabilities of application of projection methods, having proposed not necessarily just sharing the same coordinate functions [5] for the series expansion and variation of the solution.

In the second step of the variational and projection methods, the approximate representation of the solution sought in the form of a series expansion in the coordinate functions is substituted in the projection conditions.

As a result of this substitution the original functional equation is approximately replaced with a finite system of linear algebraic equations whose solution uses known methods. Let us proceed to the more specific description of the variational and projection methods as applied to the problems of oscillations of an elastic continuous medium.

4.1 Variational Principle in the Problem of Forced Harmonic Vibrations for the Displacement Equation

In the operator form, an equation with respect to the amplitude of forced harmonic oscillations $\eta = \eta(x)$ with a frequency λ has the form (1.75)

$$N\eta - \lambda^2 R\eta - f = 0, \quad N = DB^{-1}D^*.$$

Let us assume in what follows that the boundary conditions on displacements and stresses (internal forces) are simply homogeneous. This means that the operators D and D^* are simply conjugate only, while the operator N will be self-conjugate and positive definite. Exactly the same properties are found in the inertia operator R.

Let us introduce the functional

$$l(\eta) = \frac{1}{2}\int_{\Omega} (N\eta - \lambda^2 R\eta - 2f)\eta \mathrm{d}\Omega \tag{4.1}$$

and prove that the above equation on displacement is equivalent to the condition of stationarity of this functional (i.e. to the condition that variation of the functional can be represented by $\delta l(\eta) = 0$).

Really, let us give the variation, that is an infinitesimal change $\delta\eta(x)$, to the displacement $\eta(x)$ in expression (4.1). We assume that the function $\delta\eta(x)$ also belongs to the domain of definition of the operator N, including fulfillment of simple homogeneous boundary conditions. Let us calculate the corresponding variation of the functional $\delta l(\eta)$

$$\delta l(\eta) = \frac{1}{2}\int_{\Omega} (N\delta\eta - \lambda^2 R\delta\eta)\eta d\Omega + \frac{1}{2}\int_{\Omega} (N\eta - \lambda^2 R\eta - 2f)\delta\eta d\Omega.$$

In view of self-conjugacy, or self-adjointness, of the operators N and R

$$\delta l(\eta) = \frac{1}{2}\int_{\Omega} (\delta\eta N\eta - \lambda^2 \delta\eta R\eta) d\Omega$$
$$+ \frac{1}{2}\int_{\Omega} (N\eta - \lambda^2 R\eta - 2f)\delta\eta d\Omega = \int_{\Omega} (N\eta - \lambda^2 R\eta - f)\delta\eta d\Omega.$$

The condition of stationarity of the functional $\delta l(\eta) = 0$ can be written as

$$\int_{\Omega} (N\eta - \lambda^2 R\eta - f)\delta\eta d\Omega = 0 \tag{4.2}$$

or

$$\int_{\Omega} \Delta\delta\eta d\Omega = 0, \tag{4.3}$$

where the residual

$$\Delta = \Delta(x) = N\eta - \lambda^2 R\eta - f.$$

Obviously, if Eq. (1.75) holds (i.e. $\Delta = 0$), then condition (4.3) is true. Conversely, $\Delta = 0$ by virtue of equality (4.3) and the arbitrariness of variation $\delta\eta$. The latter statement can be formulated in another way: the load $\Delta(x)$, which works on any virtual (possible) displacement $\delta\eta(x)$, is equal to zero and itself vanishes identically (i.e. is identically equal to zero); in short, the principle of virtual displacements in mechanics (the virtual work principle).

4.2 Variational Principle in the Problem of Free Harmonic Vibrations for the Equation on Displacements Using a Differential Operator

From the equation regarding the amplitude of free oscillations (3.4)

$$N\eta - \lambda^2 R\eta = 0$$

it follows that

$$\lambda^2(\eta) = \frac{\int_\Omega N\eta\eta d\Omega}{\int_\Omega R\eta\eta d\Omega}, \tag{4.4}$$

in which connection $\int_\Omega R\eta\eta d\Omega > 0$ at $\eta(x) \neq 0$ by virtue of the positive definiteness of the operator R.

Expression (4.4) can be regarded as a functional, whose argument is the function $\eta(x)$, belonging to the domain of definition of the operator N. Such a function is nothing more than virtual (possible) displacement.

The variational principle: Eq. (3.4) is equivalent to the condition of stationarity of the functional (4.4). In other words, the forms of free oscillations give a stationary value to the functional (4.4). Hence, the magnitude of the functional will be equal to the square of the appropriate frequency of free oscillations.

Let us prove this. Calculate the variation of the functional (4.4)

$$\begin{aligned}\delta\lambda^2(\eta) &= \frac{2\int_\Omega N\eta\delta\eta d\Omega \int_\Omega R\eta\eta d\Omega - 2\int_\Omega N\eta\eta d\Omega \int_\Omega R\eta\delta\eta d\Omega}{\left(\int_\Omega R\eta\eta d\Omega\right)^2} \\ &= \frac{2}{\int_\Omega R\eta\eta d\Omega} \int\limits_\Omega (N\eta - \lambda^2 R\eta)\delta\eta d\Omega.\end{aligned}$$

By doing so, the condition of stationarity of the functional $\lambda^2(\eta)$ will be

$$\int\limits_\Omega (N\eta - \lambda^2 R\eta)\delta\eta d\Omega = 0. \tag{4.5}$$

Obviously, the function $\eta(x)$ that satisfies Eq. (3.4) (i.e. a form of free oscillations) gives the zero value to the integral (4.5) and to the variation $\delta\lambda^2(\eta)$. Conversely, from condition (4.5) and the arbitrariness of variation $\delta\eta$, Eq. (3.4) holds. Comparing (4.4) and (3.7), we see that the value of the functional (4.4) is equal to the square of the frequency of free oscillations.

Note that the conditions of stationarity in the problem of forced and free oscillations (4.2) and (4.5) have a unified view, but in the second case the load is $f = 0$. The consequence of the above is the assertion that the function $\eta(x)$, which gives the functional (4.4) the minimum value, is the first form of free oscillations and the corresponding value of the functional (4.4) is equal to the square of the first frequency of free oscillations. This means the minimum of the functional is characterized by a condition in which its variation vanishes. At the same time, as has just been shown, the function $\eta(x)$ represents one of the forms of free oscillations $\eta_i(x)$, while the functional (4.4) takes on the value of squared frequency of free oscillations λ_i^2. The condition means the functional is minimal and, therefore, equal to $\min \lambda_i^2 (i = 1, 2, \ldots) = \lambda_1^2$, so $\eta(x) = \eta_1(x)$. Thus, in the problem of free oscillations for any forms and frequencies of free oscillations the variational principle of stationarity can be formulated, while for the first form and frequency of oscillations this applies to the principle of minimum as well.

4.3 Extreme Variational Principle in the Problem of Forced Harmonic Oscillations

Let us return to consideration of forced harmonic oscillations. Equation (1.75) is equivalent to the condition of stationarity of the functional (4.1). In addition, at $\lambda^2 < \lambda_1^2$ this functional not only has a stationary, but also a minimal value. The above is particularly relevant to statics when $\lambda^2 = 0$.

To prove this, we calculate the second variation of the functional (4.1)

$$\delta^2 I(\eta) = \int\limits_\Omega (N - \lambda^2 R)\delta\eta\delta\eta d\Omega$$
$$= \int\limits_\Omega R\delta\eta\delta\eta d\Omega \left(\frac{\int_\Omega N\delta\eta\delta\eta d\Omega}{\int_\Omega R\delta\eta\delta\eta d\Omega} - \lambda^2 \right).$$

According to the foregoing

$$\min \frac{\int_\Omega N\delta\eta\delta\eta d\Omega}{\int_\Omega R\delta\eta\delta\eta d\Omega} = \lambda_1^2,$$

from which

$$\delta^2 I(\eta) > \int\limits_\Omega R\delta\eta\delta\eta d\Omega \cdot (\lambda_1^2 - \lambda^2),$$

and the second variation of the functional (4.1) is positive, if

$$\lambda^2 < \lambda_1^2. \tag{4.6}$$

Hence, everywhere inside the frequency zone up to the first resonance the principle of minimum of the functional (4.1) holds, while in the general case when (4.6) is not satisfied only the principle of stationarity is valid.

In statics, when $\lambda^2 = 0$

$$l(\eta) = \frac{1}{2}\int\limits_{\Omega} (N\eta - 2f)\eta \mathrm{d}\Omega = \frac{1}{2}\int\limits_{\Omega} N\eta\eta \mathrm{d}\Omega - \int\limits_{\Omega} f\eta \mathrm{d}\Omega.$$

The first integral represents the work of the forces of elasticity

$$\begin{aligned}\frac{1}{2}\int\limits_{\Omega} N\eta\eta \mathrm{d}\Omega &= \frac{1}{2}\int\limits_{\Omega} DB^{-1}D^{*}\eta\eta \mathrm{d}\Omega \\ &= \frac{1}{2}\int\limits_{\Omega} B^{-1}D^{*}\eta D^{*}\eta \mathrm{d}\Omega = \frac{1}{2}\int\limits_{\Omega} \xi\varepsilon \mathrm{d}\Omega.\end{aligned}$$

The second integral represents doubling the work of the external load, such that there will be

$$l(\eta) = \frac{1}{2}\int\limits_{\Omega} \xi\,\varepsilon \mathrm{d}\Omega - 2\cdot\frac{1}{2}\int\limits_{\Omega} f\eta \mathrm{d}\Omega.$$

The principle of minimum of this functional for actual (real) displacements in statics is known [3].

Under condition (4.6) exactly the same principle is valid for harmonic preresonant vibrations as well.

4.4 Mixed Variational Principle in the Problem of Forced Harmonic Oscillations (Principle of Reissner)

Let us rewrite the system of equations of the form (1.35)

$$\begin{aligned} &D\xi - \lambda^2 R\eta - f = 0, \\ &D^{*}\eta - B\xi - e = 0 \end{aligned} \tag{4.7}$$

with homogeneous boundary conditions and compose a functional that depends on two variables ξ and η, characterizing the stress state and displacement

$$l(\xi, \eta) = \frac{1}{2}\int\limits_{\Omega} (D\xi - \lambda^2 R\eta - 2f)\eta \mathrm{d}\Omega + \frac{1}{2}\int\limits_{\Omega} \xi(D^*\eta - B\xi - 2e)\mathrm{d}\Omega. \tag{4.8}$$

The mixed variational principle can be formulated by making the system of Eq. (4.7) equivalent to the condition of stationarity of the functional (4.8).

The principle was first proposed in [6].

This can be proved by putting together an expression for the variation

$$\delta l(\xi, \eta) = \frac{1}{2}\int\limits_{\Omega} (D\delta\xi - \lambda^2 R\delta\eta)\eta \mathrm{d}\Omega + \frac{1}{2}\int\limits_{\Omega} (D\xi - \lambda^2 R\eta - 2f)\delta\eta \mathrm{d}\Omega + \frac{1}{2}\int\limits_{\Omega} \xi(D^*\delta\eta - B\delta\xi)\mathrm{d}\Omega + \frac{1}{2}\int\limits_{\Omega} \delta\xi(D^*\eta - B\xi - 2e)\mathrm{d}\Omega.$$

We have shown that the operators D and D^* are conjugate and the operators R and B are self-conjugate. Hence

$$\int\limits_{\Omega} D\delta\xi\, \eta \mathrm{d}\Omega = \int\limits_{\Omega} \delta\xi D^*\eta \mathrm{d}\Omega, \quad \int\limits_{\Omega} R\delta\eta\eta \mathrm{d}\Omega = \int\limits_{\Omega} R\eta\delta\eta \mathrm{d}\Omega,$$

$$\int\limits_{\Omega} B\delta\xi\xi \mathrm{d}\Omega = \int\limits_{\Omega} B\xi\delta\xi \mathrm{d}\Omega.$$

Then there will be

$$\delta l(\xi, \eta) = \int\limits_{\Omega} (D\xi - \lambda^2 R\eta - f)\delta\eta \mathrm{d}\Omega + \int\limits_{\Omega} \delta\xi(D^*\eta - B\xi - e)\mathrm{d}\Omega = 0.$$

By virtue of independence of the variations $\delta\xi$ and $\delta\,\eta$

$$\int\limits_{\Omega}(D\xi-\lambda^{2}R\eta-f)\delta\eta\mathrm{d}\Omega=0,$$
$$\int\limits_{\Omega}\delta\xi(D^{*}\eta-B\xi-e)\mathrm{d}\Omega=0. \tag{4.9}$$

These conditions are obviously satisfied if the equations of the form (4.7) are valid. The inverse follows from arbitrariness of the variations $\delta\xi$ and $\delta\,\eta$.

4.5 Variational Principle in the Problem of Forced Harmonic Vibrations for the Equation on Displacement Using an Integral Operator

The equation with an integral operator (1.112) is

$$\eta-\lambda^{2}AR\eta-\varphi=0,$$

where $\varphi=Af$ is equivalent to the condition of stationarity of the functional

$$l(\eta)=\frac{1}{2}\int\limits_{\Omega}R\eta(\eta-\lambda^{2}AR\eta-2\varphi)\mathrm{d}\Omega. \tag{4.10}$$

This is easily verified, having written an expression for variation of the functional which can be recorded as follows

$$\delta l(\eta)=\int\limits_{\Omega}\delta(R\eta)(\eta-\lambda^{2}AR\eta-\varphi)\mathrm{d}\Omega$$

where $R\,\eta$ represents the inertial load $\lambda^{2}R\eta$ at $\lambda^{2}=1$.

It turns out that variation of the functional is the work of the load $\delta(R\eta)$ on the displacement $(\eta-\lambda^{2}AR\eta-\varphi)$.

4.6 Variational Principle in the Problem of Nonstationary Vibrations

To the equation of nonstationary oscillations (1.74)

$$N\eta + R\ddot{\eta} - f = 0$$

we accordingly associate the functional

$$l(\eta) = \frac{1}{2}\int_0^t \mathrm{d}\tau \int_\Omega ((N\eta - 2f)\eta - R\dot{\eta}\dot{\eta})\mathrm{d}\Omega, \tag{4.11}$$

where t, τ is time.

Let us calculate the variation of the functional (4.11)

$$\delta l(\eta) = \frac{1}{2}\int_0^t \mathrm{d}\tau \int_\Omega 2((N\eta - f)\delta\eta - R\dot{\eta}\delta\dot{\eta})\mathrm{d}\Omega,$$

where $\delta\eta = \delta\eta(x,\tau)$ assuming that $\delta\eta(x,0) = 0, \delta\eta(x,t) = 0$. Hence, there will be

$$\int_0^t \mathrm{d}\tau \int_\Omega R\dot{\eta}\delta\dot{\eta}\mathrm{d}\Omega = \int_\Omega R\dot{\eta}\delta\eta|_0^t \mathrm{d}\Omega - \int_0^t \mathrm{d}\tau \int_\Omega R\ddot{\eta}\delta\eta\mathrm{d}\Omega$$
$$= -\int_0^t \mathrm{d}\tau \int_\Omega R\ddot{\eta}\delta\eta\mathrm{d}\Omega.$$

The resulting condition of stationarity

$$\delta l(\eta) = \int_0^t \int_\Omega (N\eta + R\ddot{\eta} - f)\,\delta\eta\mathrm{d}\Omega\mathrm{d}\tau = 0$$

at an arbitrary $\delta\eta(x,\tau)$, where x belongs to the domain Ω and $0 \le \tau \le t$ is equivalent to the equation of the form (1.74).

4.7 Rheonomic Variation

Let us consider the functional

$$l(\eta) = \frac{1}{2}\int\limits_{\Omega} (N\eta + 2(R\ddot{\eta} - f))\eta \mathrm{d}\Omega, \tag{4.12}$$

depending on time t as a parameter. Let us give a variation $\delta\eta(x)$ that does not depend on time t (rheonomic variation) to displacement $\eta(x, t)$. This means the variation of the functional

$$\begin{aligned}\delta l(\eta) &= \frac{1}{2}\int\limits_{\Omega} (2N\eta + 2(R\ddot{\eta} - f))\delta\eta(x)\mathrm{d}\Omega \\ &= \int\limits_{\Omega} (N\eta + R\ddot{\eta} - f)\delta\eta(x)\mathrm{d}\Omega,\end{aligned}$$

since $\delta(\ddot{\eta}) = (\delta\eta)^{\cdot\cdot} = 0$. By the arbitrariness of variation $\delta\,\eta\,(x)$ the stationarity condition $\delta l(\eta) = 0$ and equation of the form (1.74) are equivalent at any given time. We note that the variational principle, based on rheonomic variation, also extends to the case when viscous friction forces act as well as the forces of inertia, elasticity and external load. In what follows G is the algebraic operator of friction.

This means there will be

$$N\eta + R\ddot{\eta} + G\dot{\eta} - f = 0,$$

$$l(\eta) = \frac{1}{2}\int\limits_{\Omega} (N\eta + 2(R\ddot{\eta} + G\dot{\eta} - f))\eta \mathrm{d}\Omega,$$

$$\delta l(\eta) = \int\limits_{\Omega} (N\eta + R\ddot{\eta} + G\dot{\eta} - f)\delta\eta(x)\mathrm{d}\Omega = 0$$

since also $\delta(\dot{\eta}) = (\delta\eta)^{\cdot} = 0$.

4.8 Variational Method for Solving Equations of the Theory of Vibrations in Displacements

To construct an algorithm for an approximate method, let us take advantage of the corresponding variational condition. In the problem of forced harmonic oscillations, this condition has the form (4.2)

$$\int_{\Omega} (N\eta - \lambda^2 R\eta - f)\delta\eta \mathrm{d}\Omega = 0.$$

We choose a sequence of functions $\eta_i = \eta_i(x), i = 1, 2, \ldots, n$, such that

(1) $\eta_i(x)$ belong to the domain of definition of N;
(2) functions $\eta_i(x)$ are linearly independent;
(3) the sequence of functions $\eta_i(x), i = 1, 2, \ldots, n$, possesses the completeness necessary and sufficient for approximation of an arbitrary displacement $\eta(x)$.

The amplitude of displacement $\eta(x)$ sought can be represented as a linear combination of functions $\eta_i(x), i = 1, 2, \ldots, n$. However, to obtain the approximate solution we confine ourselves to a finite number of members of the series

$$\eta = \sum_{i=1}^{n} \beta_i \eta_i. \tag{4.13}$$

The functions $\eta_i(x), i = 1, 2, \ldots, n$, are referred to as coordinate functions. This means the variation of displacement is

$$\delta\eta = \sum_{i=1}^{n} \delta\beta_i \eta_i, \tag{4.14}$$

where $\delta\beta_i, i = 1, 2, \ldots, n$, are arbitrary variations of the coefficients of the series (4.13).

Let us substitute expression (4.14) into equality (4.2)

$$\int_{\Omega} (N\eta - \lambda^2 R\eta - f) \sum_{i=1}^{n} \delta\beta_i \eta_i \mathrm{d}\Omega = 0.$$

Reversing the order of integration and summation, we obtain

$$\sum_{i=1}^{n} \delta\beta_i \int_{\Omega} (N\eta - \lambda^2 R\eta - f)\eta_i \mathrm{d}\Omega = 0.$$

The latter relationship can take place only if each of the integrals is equal to zero. This statement as a consequence is subject to projection conditions

$$\int_{\Omega} (N\eta - \lambda^2 R\eta - f)\eta_i \mathrm{d}\Omega = 0, \quad i = 1, 2, \ldots, n, \tag{4.15}$$

or

$$\int_{\Omega} \Delta \eta_i \mathrm{d}\Omega = 0, \quad i = 1, 2, \ldots, n, \tag{4.16}$$

where $\Delta = N\eta - \lambda^2 R\eta - f$.

A geometric interpretation can be given to the projection conditions.

The equation of the form (1.75) and conditions (4.15) can be rewritten as follows

$$\Delta_0 = 0 \quad \text{and} \quad \int_{\Omega} \Delta_0 \eta_{0i} \mathrm{d}\Omega = 0, \; i = 1, 2, \ldots, n, \tag{4.17}$$

where $\eta_{0i} = \sqrt{R}\eta_i, \Delta_0 = (\sqrt{R})^{-1}\Delta$.

Let us assume that functions η_{0i} are not only linearly independent, but also orthonormal. Hence, the transition from one system of the functions to the other can be carried out using the procedure of orthonormality of Schmidt [7].

In the n-dimensional space of a Cartesian coordinate system with unit vectors η_{0i}, there is a condition similar to (4.17)

$$(\Delta_0, \eta_{0i}) = 0, \quad i = 1, 2, \ldots, n, \tag{4.18}$$

which represents the well-known fact that the n-dimensional vector $\Delta_0 = 0$ if and only if all its projections are zeros.

Another equation can be given to the projection Eq. (4.15), recollecting that $N = DB^{-1}D^*$

$$\begin{aligned} \int_{\Omega} N\eta\eta_i \mathrm{d}\Omega = \int_{\Omega} DB^{-1}D^*\eta\eta_i \mathrm{d}\Omega = -\int_{\Gamma} B^{-1}D^*\eta\eta_i \mathrm{d}\Gamma + \int_{\Omega} B^{-1}D^*\eta D^*\eta_i \mathrm{d}\Omega \\ = -\int_{\Gamma} \xi\eta_i \mathrm{d}\Gamma + \int_{\Omega} B^{-1}D^*\eta D^*\eta_i \mathrm{d}\Omega. \end{aligned} \tag{4.19}$$

For the simple homogeneous boundary conditions given on the forces and displacements, let us facilitate the choice of coordinate functions η_i, since they belong to the domain of definition of the operator D^* and satisfy boundary conditions only on displacements. This gives us the possibility to calculate the integral and to turn the integrated part in Eq. (4.19) into zero.

We obtain

$$\int_{\Omega} N\eta\eta_i \mathrm{d}\Omega = \int_{\Omega} B^{-1}D^*\eta D^*\eta_i \mathrm{d}\Omega, \tag{4.20}$$

and instead of (4.15) we have

$$\int_{\Omega} B^{-1} D^* \eta D^* \eta_i d\Omega - \int_{\Omega} (\lambda^2 R\eta + f)\eta_i d\Omega = 0. \tag{4.21}$$

Let us substitute expansion (4.13) into formulas (4.15)

$$\int_{\Omega} \left(N \sum_{j=1}^{n} \beta_j \eta_j - \lambda^2 R \sum_{j=1}^{n} \beta_j \eta_j - f \right) \eta_i d\Omega = 0, \quad i = 1, 2, \ldots, n.$$

Interchanging the order of integration and summation, we arrive at the system of n equations for n coefficients of expansion (4.13)

$$\sum_{j=1}^{n} N_{ji} \beta_j - \lambda^2 \sum_{j=1}^{n} R_{ji} \beta_j - f_i = 0, \quad i = 1, 2, \ldots, n, \tag{4.22}$$

where the following notations have been introduced

$$N_{ji} = \int_{\Omega} N \eta_j \eta_i d\Omega,$$

$$R_{ji} = \int_{\Omega} R \eta_j \eta_i d\Omega, \quad f_i = \int_{\Omega} f \eta_i d\Omega.$$

Having found the coefficients $\beta_i, i = 1, 2, \ldots, n$, from the system of Eqs. (4.22) using formula (4.13), we construct a solution to the problem. In the case of free vibrations, when there is no exciting load, the variation condition takes on the form of (4.5)

$$\int_{\Omega} (N\eta - \lambda^2 R\eta) \delta\eta d\Omega = 0.$$

As a result of substituting (4.13) and (4.14), we obtain

$$\sum_{j=1}^{n} N_{ji} \beta_j - \lambda^2 \sum_{j=1}^{n} R_{ji} \beta_j = 0, \quad i = 1, 2, \ldots, n. \tag{4.23}$$

The system of homogeneous linear algebraic equations (4.23) regarding the coefficients $\beta_i, i = 1, 2, \ldots, n$, is solvable, if the determinant

$$|N_{ji} - \lambda^2 R_{ji}| = 0; \quad j, i = 1, 2, \ldots, n.$$

The equation concerning λ^2 has n roots $\lambda_k^2, k = 1, 2, \ldots, n$, where λ_k is the kth approximate frequency of free oscillations. Each frequency has its own solution to the system of algebraic equations (4.23) and, by virtue of (4.13), its own appropriate approximate form of free oscillations.

It follows from the projection equalities (4.21) that in Eqs. (4.22), (4.23) there will be

$$N_{ji} = \int_{\Omega} B^{-1} D^* \eta_j D^* \eta_i \mathrm{d}\Omega. \tag{4.24}$$

4.9 Approximate Method of Solving the Problem of Harmonic Vibrations, Based on the Mixed Variational Principle

Let us write the variational conditions (4.9)

$$\int_{\Omega} (D\xi - \lambda^2 R\eta - f)\delta\eta \mathrm{d}\Omega = 0,$$

$$\int_{\Omega} \delta\xi (D^*\eta - B\xi - e)\mathrm{d}\Omega = 0.$$

Let us expand the stress state and displacement independently

$$\xi = \sum_{i=1}^{m} \alpha_i \xi_i, \quad \eta = \sum_{i=1}^{n} \beta_i \eta_i. \tag{4.25}$$

We assume that the coordinate functions ξ_i and η_i belong to the domain of definition of the operators D and D^*, respectively: $D\xi_i = h_i$, $D^*\eta_i = \varepsilon_i$.

The functions $\xi_i(x)$ are linearly independent, while the sequence of functions $\xi_i(x)$, $i = 1, 2, \ldots, m$, possesses the completeness necessary and sufficient for approximation of an arbitrary stress state $\xi(x)$. A similar assumption is adopted concerning the coordinate functions $\eta_i(x)$ and approximation of an arbitrary displacement $\eta(x)$ by means of the sequence of functions $\eta_i(x)$, $i = 1, 2, \ldots, n$.

There are variations

$$\delta\xi = \sum_{i=1}^{m} \delta\alpha_i \xi_i, \quad \delta\eta = \sum_{i=1}^{n} \delta\beta_i \eta_i. \tag{4.26}$$

Substitute (4.26) into the variational conditions (4.9)

$$\begin{aligned} &\int_{\Omega} (D\xi - \lambda^2 R\eta - f) \sum_{i=1}^{n} \delta\beta_i \eta_i \mathrm{d}\Omega = 0, \\ &\int_{\Omega} \sum_{i=1}^{m} \delta\alpha_i \xi_i (D^* \eta - B\xi - e) \mathrm{d}\Omega = 0. \end{aligned} \tag{4.27}$$

Interchanging the order of integration and summation, as a result of the arbitrariness of variations $\delta\alpha_i$ and $\delta\beta_i$ we obtain the projection equalities

$$\begin{aligned} &\int_{\Omega} (D\xi - \lambda^2 R\eta - f)\eta_i \mathrm{d}\Omega = 0, \quad i = 1, 2, \ldots, n, \\ &\int_{\Omega} \xi_i (D^*\eta - B\xi - e) \mathrm{d}\Omega = 0, \quad i = 1.2, \ldots, m. \end{aligned} \tag{4.28}$$

Let us substitute the series (4.25) into the projection equalities (4.28). We will come to the system of $n+m$ linear algebraic equations for the coefficients $\alpha_{,i}$, $i = 1, 2, \ldots, m$, and $\beta_i, i = 1, 2, \ldots, n$

$$\begin{aligned} &\sum_{j=1}^{m} D_{ji}\alpha_j - \lambda^2 \sum_{j=1}^{n} R_{ji}\beta_j - f_i = 0, \quad i = 1, 2, \ldots, n, \\ &\sum_{j=1}^{n} D^*_{ij}\beta_j - \sum_{j=1}^{m} B_{ij}\alpha_j - e_i = 0, \quad i = 1, 2, \ldots, m, \end{aligned}$$

where it is designated that

$$\begin{aligned} &D_{ji} = \int_{\Omega} D\xi_j \eta_i \mathrm{d}\Omega, \quad D^*_{ij} = \int_{\Omega} \xi_i D^* \eta_j \mathrm{d}\Omega, \quad R_{ji} = \int_{\Omega} R\eta_j \eta_i \mathrm{d}\Omega, \\ &B_{ij} = \int_{\Omega} B\xi_i \xi_j \mathrm{d}\Omega, \quad f_i = \int_{\Omega} f\eta_i \mathrm{d}\Omega, \quad e_i = \int_{\Omega} \xi_i e \mathrm{d}\Omega. \end{aligned}$$

The coefficients D_{ji} and D^*_{ij} are equal to each other at the same indexes since the operators D and D^* are conjugate.

In particular, to solve the system of Eq. (4.7) we can use eigenfunctions (forms of free oscillations) as the coordinate functions ξ_i, η_i. These are taken from the simpler problem

$$D\xi_i - \lambda_i^2 R_0 \eta_i = 0,$$
$$D^* \eta_i - B_0 \xi_i = 0,$$

where R_0, B_0 are the unit operators of inertia and elasticity, even while retaining homogeneous boundary conditions.

It is possible to replace the coordinate function ξ_i in formula (4.26) by the coordinate function $B^{-1}B_0\xi_i$. Then, under $e = 0$ the result gotten for displacements coincides with the one given by formulas (4.22) and (4.24). Such a variant of the mixed variational method can be treated as a Petrov-Galerkin method.

4.10 Formulas of the First Approximation for the First Frequency of Free Oscillations (Formulas of Rayleigh and Ritz)

Let us confine ourselves to a single member in the series expansion (4.13) of the solution sought to the problem of free oscillations in coordinate functions: $\eta = \beta_1 \eta_1$, where $\eta = \eta(x), \eta_1 = \eta_1(x)$. What we need to find is λ_1^2.

Hence, the system of Eq. (4.23) degenerates into the single equation

$$N_{11}\beta_1 - \lambda_1^2 R_{11}\beta_1 = 0,$$

from which

$$\lambda_1^2 = \frac{N_{11}}{R_{11}} = \frac{\int_\Omega N\eta_1\eta_1 \mathrm{d}\Omega}{\int_\Omega R\eta_1\eta_1 \mathrm{d}\Omega}. \tag{4.29}$$

Let us denote $N\eta_1 = h_1$, such that $\eta_1 = Ah_1$ and

$$\lambda_1^2 = \frac{\int_\Omega h_1\eta_1 \mathrm{d}\Omega}{\int_\Omega R\eta_1\eta_1 \mathrm{d}\Omega}. \tag{4.30}$$

This approximate expression for the first natural frequency of oscillations is simply a generalization of the formula of Rayleigh, who proposed accepting the static deflection under the action of the weight load $h_1(x)$ as the approximate first natural form of oscillations $\eta_1(x)$.

The Rayleigh method is also called the "energy method". According to (4.30), this means

$$\frac{1}{2}\lambda_1^2 \int_\Omega R\eta_1\eta_1 \mathrm{d}\Omega = \frac{1}{2}\int_\Omega h_1\eta_1 \mathrm{d}\Omega. \tag{4.31}$$

The maximum kinetic energy is on the left-hand side of this equality, while the work of load $h_1(x)$ on displacement $\eta_1(x)$ is on the right-hand side. With homogeneous boundary conditions on the forces and displacements, this load is equal to the maximal potential energy of deformation by virtue of conjugacy of the operators D and D^*.

Therefore

$$\begin{aligned}\frac{1}{2}\int\limits_{\Omega} h_1\eta_1 \mathrm{d}\Omega = \frac{1}{2}\int\limits_{\Omega} N\eta_1\eta_1 \mathrm{d}\Omega = \frac{1}{2}\int\limits_{\Omega} DB^{-1}D^*\eta_1\eta_1 \mathrm{d}\Omega \\ = \frac{1}{2}\int\limits_{\Omega} B^{-1}D^*\eta_1 D^*\eta_1 \mathrm{d}\Omega = \frac{1}{2}\int\limits_{\Omega} B^{-1}\varepsilon_1\varepsilon_1 \mathrm{d}\Omega,\end{aligned} \tag{4.32}$$

where $\varepsilon_1 = D^*\eta_1$; and $\varepsilon_1(x)$ is the strain corresponding to the displacement $\eta_1(x)$.

Hence, instead of (4.30) we can write the formula of Ritz [8]

$$\lambda_1^2 = \frac{\int_\Omega B^{-1}D^*\eta_1 D^*\eta_1 \mathrm{d}\Omega}{\int_\Omega R\eta_1\eta_1 \mathrm{d}\Omega}. \tag{4.33}$$

Using the Ritz formula (4.33) instead of the Rayleigh formula (4.30) facilitates selection of the coordinate function $\eta_1(x)$

$$D^*\eta_1 = \varepsilon_1 \tag{4.34}$$

since there is a need to implement boundary conditions on the displacements. However, according to Rayleigh it is still necessary that $\eta_1(x)$ belongs to the domain of definition of the operator N

$$N\eta_1 = h_1, \tag{4.35}$$

which means the boundary conditions at both displacements and forces are satisfied. Therefore, the Ritz formula (4.33) is, generally speaking, less accurate than the Rayleigh formula (4.30). Note that formulas (4.30) and (4.33) give one and the same result at performance (4.35).

We make use of the Ritz formula for approximate calculation of the first natural frequency of transverse vibrations of a beam, simply supported at the ends on rigid supports. Let mass per unit length of the beam be $\rho(x)$ and bending compliance $\beta(x) = \frac{1}{EJ(x)}$, where E is the modulus of elasticity, $J(x)$ the moment of cross-section inertia. In the problem considered the operators are $R = \rho, B^{-1} = \frac{1}{\beta} = EJ, \quad D^* = -\frac{\mathrm{d}^2}{\mathrm{d}x^2}$.

The boundary conditions on the displacements are of the form

$$\eta(0) = \eta(l) = 0. \tag{4.36}$$

Let us choose the coordinate function

$$\eta_1(x) = \sin\frac{\pi x}{l}. \tag{4.37}$$

Obviously, $\eta_1(x)$ belongs to the domain of definition of the operator D^*

$$D^*\eta_1 = -\frac{\mathrm{d}^2}{\mathrm{d}x^2}\eta_1 = \left(\frac{\pi}{l}\right)^2 \sin\frac{\pi x}{l},$$

and boundary conditions (4.36) are satisfied.

The Ritz formula (4.33) for the first frequency of free oscillations of a beam, simply supported at the ends on rigid supports, will be written as

$$\lambda_1^2 = \left(\frac{\pi}{l}\right)^4 \frac{\int_0^l EJ \sin^2\frac{\pi x}{l}\mathrm{d}x}{\int_0^l \rho \sin^2\frac{\pi x}{l}\mathrm{d}x}. \tag{4.38}$$

This formula enables us to calculate approximately the first frequency of free oscillations of a beam of variable cross-section, while the selected coordinate function (4.37) is the first form of free oscillations of a beam of constant section.

Mixed variational principle: the system of algebraic equations in the problem of free oscillations in the first approximation will be

$$D_{11}\alpha_1 - \lambda_1^2 R_{11}\beta_1 = 0,$$
$$D_{11}^*\beta_1 - B_{11}\alpha_1 = 0.$$

The condition of solvability of this homogeneous system is

$$D_{11}D_{11}^* - \lambda_1^2 R_{11}B_{11} = 0,$$

from which we get for the first frequency of free oscillations

$$\lambda_1^2 = \frac{D_{11}D_{11}^*}{R_{11}B_{11}}. \tag{4.39}$$

Let us choose the functions

$$\xi_1(x) = \sin\frac{\pi x}{l}, \quad \eta_1(x) = \sin\frac{\pi x}{l}$$

as coordinate functions for approximation of both the bending moment and displacement.

These functions are selected to calculate the following derivatives

$$D^*\eta_1 = -\frac{d^2}{dx^2}\left(\sin\frac{\pi x}{l}\right) = \left(\frac{\pi}{l}\right)^2 \sin\frac{\pi x}{l},$$
$$D\xi_1 = -\frac{d^2}{dx^2}\left(\sin\frac{\pi x}{l}\right) = \left(\frac{\pi}{l}\right)^2 \sin\frac{\pi x}{l}.$$

The coordinate functions selected also satisfy the conditions of simple support $(\eta(0) = \eta(l) = 0, \xi(0) = \xi(l) = 0)$.

Further

$$D_{11} = D_{11}^* = \int_0^l D\xi_1\eta_1 dx = \int_0^l \xi_1 D^*\eta_1 dx$$
$$= \int_0^l \left(\frac{\pi}{l}\right)^2 \sin^2\frac{\pi x}{l} dx = \left(\frac{\pi}{l}\right)^2 \frac{l}{2}$$

and

$$R_{11} = \int_0^l R\eta_1\eta_1 dx = \int_0^l \rho \sin^2\frac{\pi x}{l} dx,$$
$$B_{11} = \int_0^l B\xi_1\xi_1 dx = \int_0^l \beta \sin^2\frac{\pi x}{l} dx,$$

because $R = \rho, B = \beta$.

More information is used in the mixed variational principle than in the Ritz formula: both the displacement and bending moment are twice differentiable and satisfy the boundary conditions.

The first natural frequency according to the mixed variational principle can be calculated using formula (4.39) [1]

$$\lambda_1^2 = \left(\frac{\pi}{l}\right)^4 \frac{l^2}{4} \frac{1}{\int_0^l \rho \sin^2\frac{\pi x}{l} dx \int_0^l \beta \sin^2\frac{\pi x}{l} dx}. \quad (4.40)$$

Let us compare formulas (4.38) and (4.40). They are equal in the volume of calculations. However, a method based on the mixed variational principle gives greater accuracy since both the displacement and bending moment are approximated and all the boundary conditions are fulfilled, not just the conditions on displacements as is the case in the Ritz method.

4.11 Variational Difference Method in the Problem of Longitudinal Vibrations of a Rod (Free Oscillations)

There are many methods of approximate solutions of differential equations. Among them are the finite difference methods, which are based on the approximate replacement of differentials by small but finite differences. Let us show how finite difference equations can be formed based on the variational approaches [2]. Let us turn to the problem of free longitudinal vibrations of a rectilinear rod. We retain the same designations as in Chap. 1. The equations in the amplitudes of force $Q = Q(x)$ and displacement $u = u(x)$ at harmonic oscillations of a rod (1.14) and subject to

$$p = p(x) = 0, \quad e = e(x) = 0 \tag{4.41}$$

have the following form

$$-\frac{\mathrm{d}Q}{\mathrm{d}x} - \lambda^2 \rho u = 0, \quad \frac{\mathrm{d}u}{\mathrm{d}x} - bQ = 0. \tag{4.42}$$

The homogeneous boundary conditions according to (1.15) at $x = 0, l$ are

$$\begin{aligned} Q(x) &= 0 \quad \text{end of the rod is free,} \\ u(x) &= 0 \quad \text{end of the rod is clamped.} \end{aligned} \tag{4.43}$$

We partition the length of the rod into S equal portions. The length of the site is $\Delta = \frac{l}{S}$. In the middle of the sth section we position the point with coordinate $x_s, 1 \leq s \leq S$. Furthermore, we denote

$$x_0 = 0, \quad x_{S+1} = l. \tag{4.44}$$

Let us introduce the step functions

$$\xi_s(x) = \begin{cases} 1, & x_s \leq x \leq x_{s+1}, \quad s = 0, 1, \ldots, S, \\ 0, & 0 \leq x < x_s, x_{s+1} < x \leq l \quad s = 1, 2, \ldots, S-1, \\ 0, & x_{s+1} < x \leq l, \quad s = 0, \\ 0, & 0 \leq x < x_s, \quad s = S. \end{cases} \tag{4.45}$$

In addition

$$\eta_s(x) = \begin{cases} 1, & x_{s-1,s} \leq x \leq x_{s,s+1}, \quad s = 1, \ldots, S, \\ 0, & 0 \leq x < x_{s-1,s}, x_{s,s+1} < x \leq l, \quad s = 2, \ldots, S-1, \\ 0, & x_{s,s+1} < x \leq l, \quad s = 1, \\ 0, & 0 \leq x < x_{s-1,s}, \quad s = S, \end{cases} \tag{4.46}$$

where

$$x_{0,1} = x_0, \quad x_{S,S+1} = x_{S+1},$$
$$x_{s-1,s} = \frac{x_{s-1} + x_s}{2}, \quad s = 2, \ldots, S. \tag{4.47}$$

It is obvious that

$$\xi_s \xi_m = \begin{cases} \xi_s, & s = m, \\ 0, & s \neq m. \end{cases} \tag{4.48}$$

$$\eta_s \eta_m = \begin{cases} \eta_s, & s = m, \\ 0, & s \neq m. \end{cases} \tag{4.49}$$

We represent the approximate solution of the system of differential equations (4.42) in the form of step functions

$$Q(x) = \sum_{s=0}^{S} Q_{s,s+1} \xi_s(x), \quad u(x) = \sum_{s=1}^{S} u_s \eta_s(x). \tag{4.50}$$

The projection conditions stemming from the Reissner variational principle are as follows

$$\begin{aligned} &\int_0^l \left(-\frac{dQ}{dx} - \lambda^2 \rho u\right) \eta_s dx = 0, \quad s = 1, \ldots, S, \\ &\int_0^l \xi_s \left(\frac{du}{dx} - bQ\right) dx = 0, \qquad s = 0, 1, \ldots, S. \end{aligned} \tag{4.51}$$

Substituting the series (4.50) into (4.51) and using the orthogonality properties (4.48), (4.49), we get

$$\begin{aligned} &\int_0^l \left(-\frac{dQ}{dx}\right) \eta_s dx = -(Q_{s,s+1} - Q_{s-1,s}), \\ &\int_0^l \xi_s \frac{du}{dx} dx = u_{s+1} - u_s, \end{aligned} \tag{4.52}$$

$$\int_0^l (-\lambda^2 \rho u)\eta_s dx = -\lambda^2 m_s u_s,$$
$$\int_0^l \xi_s(-bQ)dx = -b_{s,s+1}Q_{s,s+1}, \tag{4.53}$$

where

$$m_s = \int_{x_{s-1,s}}^{x_{s,s+1}} \rho dx, \quad b_{s,s+1} = \int_{x_s}^{x_{s+1}} b dx. \tag{4.54}$$

As a result, we obtain a system of algebraic equations that approximately replaces the system of differential equations (4.42)

$$\begin{aligned} &-(Q_{s,s+1} - Q_{s-1,s}) - \lambda^2 m_s u_s = 0, \quad s = 1, \ldots, S, \\ &u_{s+1} - u_s - b_{s,s+1}Q_{s,s+1} = 0, \qquad s = 0, 1, \ldots, S. \end{aligned} \tag{4.55}$$

Thus, with the help of the variational difference procedure the problem on longitudinal vibrations of a straight rod of variable cross-section is approximately reduced to the problem of longitudinal vibrations of a mechanical system, consisting of a finite number of lumped masses interconnected by springs. Writing the second equation (4.55) twice for $s-1$ and s and substituting from these equations $Q_{s-1,s}$ and $Q_{s,s+1}$ into the first equation (4.55), we obtain the algebraic equations in displacements

$$a_{s,s-1}u_{s-1} + a_{ss}u_s + a_{s,s+1}u_{s+1} = 0, \quad s = 1, 2, \ldots, S, \tag{4.56}$$

where

$$\begin{aligned} &a_{s,s-1} = -c_{s-1,s}, \; a_{ss}(\lambda^2) = (c_{s-1,s} + c_{s,s+1}) - \lambda^2 m_s, \; a_{s,s+1} = -c_{s,s+1}, \\ &c_{s-1,s} = \frac{1}{b_{s-1,s}}, \quad c_{s,s+1} = \frac{1}{b_{s,s+1}}. \end{aligned} \tag{4.57}$$

Let us note the case of $s = 1$

$$a_{10}u_0 + a_{11}u_1 + a_{12}u_2 = 0. \tag{4.58}$$

At the second boundary condition (4.43), $u_0 = u(x_0) = 0$, there will be

$$a_{11}u_1 + a_{12}u_2 = 0. \tag{4.59}$$

While in the case of the first boundary condition (4.43)

$$Q(x_0) = Q_{01} = \frac{(u_1 - u_0)}{b_{01}} = 0,$$

and the string (4.58) will be rewritten such that

$$(a_{10} + a_{11})u_1 + a_{12}u_2 = 0. \tag{4.60}$$

In other words, at any of the boundary conditions (4.43) the first equation ($s = 1$) in the system (4.56) comprises only two summands. The same remark applies to the last equation at $s = S$ too. All the other remaining equations of the system (4.56) contain exactly three terms. That is, the matrix of this system of algebraic equations has the form

$$\begin{matrix} a_{11} & a_{12} & & & & & \\ a_{21} & a_{22} & a_{23} & & & & \\ & \cdots & \cdots & \cdots & & & \\ & & a_{s,s-1} & a_{ss} & a_{s,s+1} & & \\ & & & \cdots & \cdots & \cdots & \\ & & & & a_{S,S-1} & a_{SS} & \end{matrix} . \tag{4.61}$$

Such a matrix is called a "tridiagonal matrix".

The system of algebraic equations with a tridiagonal matrix is easily solved by Gauss's sequential elimination of unknowns. From the first equation of the type (4.59), this means that

$$u_1 = -a_{11}^{-1}a_{12}u_2 = -c_1^{-1}a_{12}u_2, \quad c_1 = a_{11}. \tag{4.62}$$

Substituting this expression for u_1 into the second equation of the system (4.56)

$$a_{21}u_1 + a_{22}u_2 + a_{23}u_3 = 0,$$

we obtain

$$c_2u_2 + a_{23}u_3 = 0,$$

where

$$c_2 = -a_{21}c_1^{-1}a_{12} + a_{22}, \tag{4.63}$$

and for any $s = 2, 3, \ldots, S - 1$

$$c_s u_s + a_{s,s+1} u_{s+1} = 0. \tag{4.64}$$

Using the recursion formula for calculation of the coefficients c_s

$$c_s = -a_{s,s-1} c_{s-1}^{-1} a_{s-1,s} + a_{ss}. \tag{4.65}$$

In the reverse direction, according to (4.64), starting from an arbitrary u_S, we have

$$u_s = -c_s^{-1} a_{s,s+1} u_{s+1}, \quad s = S-1, S-2, \ldots, 1. \tag{4.66}$$

At the end of the recurrence process at $s = S$ we obtain the equation for determining the frequencies of free oscillations

$$c_S = c_S(\lambda^2) = 0. \tag{4.67}$$

There will be exactly S such frequencies according to the number of lumped (concentrated) masses. Each frequency of free oscillations has its own form of free oscillations (i.e. the displacement vector of all masses). All forms of free oscillations can be determined up to scale since the recurrent process (4.66) begins with an arbitrary value u_S.

This way of solving a system of equations using a tridiagonal matrix is often called a "sweep method". The direct (forward) sweep is carried out by formula (4.65), while formula (4.66) determines the sequence of the reverse (backward) sweep.

4.12 Variational Difference Method (Forced Harmonic Oscillations)

The equations relative to the amplitude values of forces and displacements at forced harmonic oscillations have the form (1.14). Applying the variational difference procedure to these equations, in much the same way as we did when drawing up the homogeneous equations (4.55), we obtain

$$\begin{aligned} &-(Q_{s,s+1} - Q_{s-1,s}) - \lambda^2 m_s u_s - p_s = 0, \\ &u_{s+1} - u_s - b_{s,s+1} Q_{s,s+1} - e_{s,s+1} = 0, \end{aligned} \tag{4.68}$$

where

$$p_s = \int\limits_{x_{s-1,s}}^{x_{s,s+1}} p \, \mathrm{d}x, \quad e_{s,s+1} = \int\limits_{x_s}^{x_{s+1}} e \, \mathrm{d}x \tag{4.69}$$

and

$$\lambda = \omega \tag{4.70}$$

is the given frequency of forced oscillations, which does not coincide with any of the frequencies of free oscillations (natural frequencies). Eliminating $Q_{s-1,s}$ and $Q_{s,s+1}$ from Eq. (4.68), we obtain the equations for displacements in the case of forced harmonic oscillations

$$a_{s,s-1}u_{s-1} + a_{ss}u_s + a_{s,s+1}u_{s+1} - q_s = 0, \tag{4.71}$$

where the coefficients $a_{s,s-1}, a_{ss}(\lambda^2), a_{s,s+1}$, are still given by formulas (4.57). but at the predetermined frequency $\lambda = \omega$ and at

$$q_s = p_s + (c_{s-1,s}e_{s-1,s} - c_{s,s+1}e_{s,s+1}). \tag{4.72}$$

Let us carry out once again the Gaussian procedure of successive elimination of unknowns for Eq. (4.71), analogously to what we did in the problem of free oscillations when drawing up formulas (4.65), (4.66). Hence, there will be

$$u_1 = c_1^{-1}(-a_{12}u_2 + d_1), \quad c_1 = a_{11}, \quad d_1 = q_1.$$

Further

$$c_2u_2 + a_{23}u_3 - d_2 = 0,$$

where as before c_2 is given by formula (4.63) and

$$d_2 = q_2 - a_{21}c_1^{-1}d_1.$$

For any $s = 2, 3, \ldots, S$ there will be

$$c_su_s + a_{s,s+1}u_{s+1} - d_s = 0, \quad a_{S,S+1} = 0, \tag{4.73}$$

where the coefficient c_s is determined by the recurrence formula (4.65) and

$$d_s = q_s - a_{s,s-1}c_{s-1}^{-1}d_{s-1}. \tag{4.74}$$

Thus, direct sweeping can be carried out by formulas (4.65), (4.74) and reverse sweeping, according to (4.73), by the formulas

$$u_S = c_S^{-1}d_S, \quad u_s = -c_s^{-1}(a_{s,s+1}u_{s+1} - d_s), \; s = S-1, S-2, \ldots, 1. \tag{4.75}$$

4.13 Forced Harmonic Oscillations at the Presence of Friction

In this case, instead of (1.7), we have

$$-\frac{\partial Q}{\partial x} + g\dot{u} + \rho\ddot{u} - p = 0, \quad \frac{\partial u}{\partial x} - bQ - e = 0, \tag{4.76}$$

where $g = g(x)$ is the viscous friction coefficient.

In the same way as before (i.e. formulas (1.11))

$$p(x,t) = p(x)\cos(\lambda t + \alpha), \quad e(x,t) = e(x)\cos(\lambda t + \alpha). \tag{4.77}$$

The problems of oscillations under the action of friction forces can be solved more conveniently by moving on to the exponential functions of time. We have

$$p(x)\cos(\lambda t + \alpha) = p(x)\frac{e^{i(\lambda t+\alpha)} + e^{-i(\lambda t+\alpha)}}{2}$$

or

$$p(x,t) = \sum_{m=-1,1} p^m(x)e^{im\lambda t}, \tag{4.78}$$

and

$$e(x,t) = \sum_{m=-1,1} e^m(x)e^{im\lambda t}, \tag{4.79}$$

where

$$p^m(x) = \frac{1}{2}p(x)e^{im\alpha}, \quad e^m(x) = \frac{1}{2}e(x)e^{im\alpha}$$

Solution of the system of Eq. (4.76) can be sought by

$$u(x,t) = \sum_{m=-1,1} u^m e^{im\lambda t}, \quad Q(x,t) = \sum_{m=-1,1} Q^m e^{im\lambda t}, \tag{4.80}$$

where $u^m = (u_s^m), \quad Q^m = (Q_s^m), \quad s = 1, 2, \ldots, S$, are the column vectors.

Applying the projection procedure to Eq. (4.76) in much the same way as (4.51), substituting the representations (4.80) into the projection conditions gotten and excluding Q_s^m from the algebraic equations obtained, we get a system of algebraic equations for determining the coefficients of the series (4.80) for displacement

$$a^m_{s,s-1}u^m_{s-1} + a^m_{ss}u^m_s + a^m_{s,s+1}u^m_{s+1} - q^m_s = 0, \tag{4.81}$$

where

$$\begin{aligned} a^m_{s,s-1} &= -c_{s-1,s}, \quad a^m_{s,s+1} = -c_{s,s+1}, \\ a^m_{ss}(\lambda^2) &= (c_{s-1,s} + c_{s,s+1}) + im\lambda g_s - \lambda^2 m_s, \\ g_s &= \int\limits_{x_{s-1,s}}^{x_{s,s+1}} g\,\mathrm{d}x. \end{aligned} \tag{4.82}$$

Another form can be given to Eq. (4.81)

$$A_{s,s-1}u_{s-1} + A_{ss}u_s + A_{s,s+1}u_{s+1} - q_s = 0, \quad s = 1, \ldots, S, \tag{4.83}$$

where at $m = -1, 1$ the diagonal matrices are

$$A_{s,s-1} = \mathrm{diag}(a^m_{s,s-1}), A_{ss} = \mathrm{diag}(a^m_{s,s}), A_{s,s+1} = \mathrm{diag}(a^m_{s,s+1}), \tag{4.84}$$

and the column vectors are

$$u_s = (u^m_s), \quad q_s = (q^m_s). \tag{4.85}$$

This means the recurrence formulas (4.65), (4.74), (4.75) can be transformed into the following

$$\begin{aligned} C_s &= -A_{s,s-1}C^{-1}_{s-1}A_{s-1,s} + A_{ss}, \quad C_1 = A_{11}, \\ d_s &= q_s - A_{s,s-1}C^{-1}_{s-1}d_{s-1}, \quad d_1 = q_1, \\ u_s &= -C^{-1}_s(A_{s,s+1}u_{s+1} - C^{-1}_{ss}d_s). \end{aligned} \tag{4.86}$$

This way of solving equations like (4.83) is called the "matrix sweep method". In the recurrent formulas (4.86) the first two lines represent the direct sweep, while the last line is the reverse sweep.

4.14 Mixed Variational Principle at Complex Boundary Conditions

Let us write the equilibrium equations and elasticity relations at both static loading and harmonic oscillations of an elastic body as

$$D\xi + A\eta - f = 0, \quad D^*\eta - B\xi - e = 0, \tag{4.87}$$

where

$$A = C - \lambda^2 R, \tag{4.88}$$

C is the operator of elasticity; and λ is the frequency of oscillation.

Let us consider the case of complex boundary conditions at which

$$\begin{aligned} &X + KY - F = 0 \text{ on } \Gamma_1, \\ &Y + PX - E = 0 \text{ on } \Gamma_2, \\ &\Gamma_1 + \Gamma_2 = \Gamma, \end{aligned} \tag{4.89}$$

where Γ_1 and Γ_2 are parts of the boundary Γ of an elastic body; K and P are, respectively, the stiffness and compliance of elastic fastening; and F and E are predetermined forces and displacements on Γ_1 and Γ_2, accordingly.

The system of Eq. (4.87) and boundary conditions (4.89) are equivalent to the stationarity condition of the functional

$$\begin{aligned} l(\xi, \eta) = &\frac{1}{2}\int\limits_{\Omega} ((D\xi + A\eta - 2f)\eta + \xi(D^{*}\eta - B\xi - 2e))\mathrm{d}\Omega \\ &+ \frac{1}{2}\int\limits_{\Gamma_1} (X + KY - 2F)Y\mathrm{d}\Gamma - \frac{1}{2}\int\limits_{\Gamma_2} X(Y + PX - 2E)\mathrm{d}\Gamma. \end{aligned} \tag{4.90}$$

Hence, using the property of adjointness of the operators D and D^{*} in the sense of Lagrange and the properties of self-adjointness of the operators A and B, we can write the condition of vanishing (equality to zero) of the functional variation $\delta l(\xi, \eta)$ as

$$\begin{aligned} \delta l(\xi, \eta) = &\frac{1}{2}\int\limits_{\Omega} (D\delta\xi\eta + D\xi\delta\eta + A\delta\eta\eta + A\eta\delta\eta - 2f\delta\eta \\ &+ \delta\xi D^{*}\eta + \xi D^{*}\delta\eta - \delta\xi B\xi - \xi B\delta\xi - 2\delta\xi e)\mathrm{d}\Omega \\ &+ \frac{1}{2}\int\limits_{\Gamma_1} (X + KY - 2F)\delta Y\mathrm{d}\Gamma + \frac{1}{2}\int\limits_{\Gamma_1} (\delta X + K\delta Y)Y\mathrm{d}\Gamma \\ &- \frac{1}{2}\int\limits_{\Gamma_2} \delta X(PX + Y - 2E)\mathrm{d}\Gamma - \frac{1}{2}\int\limits_{\Gamma_2} X(P\delta X + \delta Y)\mathrm{d}\Gamma = 0 \end{aligned}$$

or

$$\delta l(\xi,\eta) = \int\limits_{\Omega} ((D\xi + A\eta - f)\delta\eta + \delta\xi(D^*\eta - B\xi - e))\mathrm{d}\Omega + \int\limits_{\Gamma_1} (X + KY - F)\delta Y \mathrm{d}\Gamma - \int\limits_{\Gamma_2} \delta X(Y + PX - E)\mathrm{d}\Gamma = 0. \tag{4.91}$$

It is evident that this equality will be a consequence of equations (4.87) and boundary conditions (4.89).

Conversely, Eq. (4.87) and boundary conditions (4.89) follow from the condition of stationarity of the functional $l(\xi,\eta)$. Let us assume that ξ and η are stress states and displacements in formula (4.91) satisfying conditions (4.89) on the boundary. Then the stationarity condition of the functional $l(\xi,\eta)$ will take the form

$$\delta l(\xi,\eta) = \int\limits_{\Omega} ((D\xi + A\eta - f)\delta\eta + \delta\xi(D^*\eta - B\xi - e))\mathrm{d}\Omega = 0. \tag{4.92}$$

However, the variations $\delta\xi$ and $\delta\eta$ in Ω are independent. Therefore, (4.92) can take place only if

$$\int\limits_{\Omega} (D\xi + A\eta - f)\delta\eta d\Omega = 0, \quad \int\limits_{\Omega} \delta\xi(D^*\eta - B\xi - e)\mathrm{d}\Omega = 0. \tag{4.93}$$

Hence, by virtue of completeness of the sets of variations $\delta\xi$ and $\delta\eta$ in the volume (domain) Ω, Eq. (4.87) follow.

If ξ and η stand for stress states and displacements in formula (4.91) that satisfy Eq. (4.87), then the stationarity condition can be written as

$$\delta l(\xi,\eta) = \int\limits_{\Gamma_1} (X + KY - F)\delta Y \mathrm{d}\Gamma - \int\limits_{\Gamma_2} \delta X(Y + PX - E)\mathrm{d}\Gamma = 0,$$

which is possible only at

$$\int\limits_{\Gamma_1} (X + KY - F)\delta Y \mathrm{d}\Gamma = 0, \quad \int\limits_{\Gamma_2} \delta X(Y + PX - E)\mathrm{d}\Gamma = 0. \tag{4.94}$$

By virtue of completeness of variation sets δY on Γ_1 and δX on Γ_2, we obtain conditions (4.89).

References

1. Fridman, V. M. (1956). On an approximate method for determining the natural frequencies of vibrations. *Academy of Sciences of the U.S.S.R., Vibrations in Turbomachinery,* pp. 69–76 (in Russian).
2. Kukishev, V. L., & Fridman, V. M. (1976). Variational difference method in the theory of elastic vibrations, based on the Reissner variational principle. *Academy of Sciences of the U.S.S.R., Mechanics of Solids*, *5*, 112–119 (in Russian).
3. Lurie, A. I. (2005). *Theory of elasticity* (1050 pp). Berlin: Springer.
4. Mikhlin, S. G. (1964). *Variational methods in mathematical physics* (p. 510). Oxford: Pergamon Press.
5. Petrov, G. I. (1940). Application of the Galerkin method to the problem of viscous fluid flow stability. *Academy of Sciences of the USSR, Applied Mathematics and Mechanics, 4*(3), 3–11 (in Russian).
6. Reissner, E. O. (1961). On some variational theorems of the theory of elasticity. *Academy of Sciences of the U.S.S.R., Problems of Continuum Mechanics,* 328–337 (in Russian).
7. Smirnov, V. I. (1964). *Course of higher mathematics* (Vol. IV, 336 pp). Oxford: Pergamon Press.
8. Ritz, W. (1909). Über eine neue Methode zur Lösung gewisser Variationsprobleme der mathematischen Physik. *Journal für die reine und angewandte Mathematik (Grelle), 135*(1), 1–61.

Chapter 5
Harmonic Analysis

Harmonic analysis refers to a way of solving the equation of oscillations, in which the solution is represented as a series expansion in the harmonic functions of time.

The operator equation in displacements has the form (1.74)

$$N\eta + R\ddot{\eta} - f = 0, \tag{5.1}$$

where, as already mentioned, $\eta = \eta(x,t)$ is the displacement and $f(x,t)$ is the external distributed load at the point of the elastic body of coordinate x at the moment of time t.

Let us first suppose that the operators N and R are not dependent of time, being simply the functions of coordinate x

$$N = N(x), \quad R = R(x).$$

As in Sect. 3.4, we consider two fundamentally different problems of the motion of an elastic body, which is described by Eq. (5.1).

(1) The problem of stationary periodic oscillations. The period of oscillations is denoted by T, $0 \le t \le T$. The conditions of periodicity are

$$\eta(x,0) = \eta(x,T), \dot{\eta}(x,0) = \dot{\eta}(x,T). \tag{5.2}$$

(2) The problem of nonstationary oscillations at given initial conditions

$$\eta(x,0) = \eta_0(x), \dot{\eta}(x,0) = v_0(x). \tag{5.3}$$

The numerical solution of the problems of linear and nonlinear oscillations at the given initial conditions (5.3) is widely carried out using the Runge-Kutta method, descriptions of which can be found in [2, 3, 5].

V. Fridman, *Theory of Elastic Oscillations*, Foundations of Engineering Mechanics, DOI 10.1007/978-981-10-4786-2_5

We mainly investigate in the present chapter periodic oscillations with conditions on time (5.2). This corresponds to the important case of the stationary operating mode of a machine. In addition, vibrations close to periodic are considered, when the oscillation process parameters change slowly in time, such as when starting and stopping the machines.

5.1 Periodic Oscillations

Periodic oscillations are sought in the form of a series of the harmonic functions of time. The harmonic functions with a period T are $\cos m\lambda t, \quad m = 0, 1, 2, 3, \ldots,$ $\sin m\lambda t, \quad m = 1, 2, 3, \ldots, \eta$ and

$$e^{im\lambda t} = \cos m\lambda t + i \sin m\lambda t, \quad m = \ldots, -3, -2, -1, 0, 1, 2, 3, \ldots,$$

where $\lambda = \frac{2\pi}{T}$ is the periodic process frequency.

It is often more convenient to use the exponential function of time $e^{im\lambda t}$, since its derivatives with respect to time $\frac{de^{im\lambda t}}{dt} = im\lambda e^{im\lambda t}$ and $\frac{d^2 e^{im\lambda t}}{dt^2} = -(m\lambda)^2 e^{im\lambda t}$ are proportional to the same function.

The important property of the functions $e^{im\lambda t}$ is their orthonormality

$$\frac{1}{T}\int\limits_0^T e^{-im\lambda t} e^{is\lambda t} dt = \delta_{ms}, \tag{5.4}$$

where δ_{ms} is the Kronecker symbol.

Suppose that the nonhomogeneous term in Eq. (5.1) is a polyharmonic function of time

$$f(x,t) = \sum_{m=-M}^{M} f^m(x) e^{im\lambda t}. \tag{5.5}$$

Let us also represent the approximate solution of Eq. (5.1) under the conditions of periodicity (5.2) in the form of the harmonic series

$$\eta(x,t) = \sum_{m=-M}^{M} \eta^m(x) e^{im\lambda t}. \tag{5.6}$$

To determine the coefficients of this series, we use the projection conditions

$$\frac{1}{T}\int_0^T \mathrm{e}^{-im\lambda t}(N\eta + R\ddot{\eta} - f)\mathrm{d}t = 0, \quad m = -M, .., M. \tag{5.7}$$

These conditions can lead to the form

$$\frac{1}{T}\int_0^T \mathrm{e}^{-im\lambda t}\Big(N\eta - (m\lambda)^2 R\eta - f\Big)\mathrm{d}t = 0, \tag{5.8}$$

integrating by parts the term containing $R\ddot{\eta}$ twice and using the periodicity condition (5.2), we get

$$\begin{aligned}\int_0^T \mathrm{e}^{-im\lambda t}R\ddot{\eta}\mathrm{d}t &= \mathrm{e}^{-im\lambda t}R\dot{\eta}\big|_{t=0}^{t=T} + im\lambda\int_0^T \mathrm{e}^{-im\lambda t}R\dot{\eta}\mathrm{d}t \\ &= im\lambda \mathrm{e}^{-im\lambda t}R\eta\big|_{t=0}^{t=T} - (m\lambda)^2\int_0^T \mathrm{e}^{-im\lambda t}R\eta\mathrm{d}t \\ &= -(m\lambda)^2\int_0^T \mathrm{e}^{-im\lambda t}R\eta\mathrm{d}t.\end{aligned}$$

Let us substitute the series (5.6) into formula (5.8), after replacing the index m in this series by s, $s = -S, \ldots, S$, $S = M$, and use the condition of orthonormality of the exponential functions (5.4). Thus, we obtain separately for the amplitude of the mth harmonic component

$$N\eta^m - (m\lambda)^2 R\eta^m - f^m = 0, \tag{5.9}$$

where

$$\eta^m = \frac{1}{T}\int_0^T \eta\,\mathrm{e}^{-im\lambda t}\mathrm{d}t, \quad f^m\frac{1}{T}\int_0^T f\mathrm{e}^{-im\lambda t}\mathrm{d}t.$$

The solutions of these equations are the coefficients of the series (5.6).

5.2 Harmonic Analysis and Spectral Method

It was shown in Chap. 3 that the amplitude of the harmonic forced vibrations of an elastic body can be represented as a series of its natural forms (i.e. the forms of its free vibrations $\eta_k(x)$, $k = 1, 2, \ldots$). Let us apply this approach to each harmonic component of polyharmonic vibrations satisfying Eq. (5.9). We can represent approximately the solution of this equation in the form

$$\eta^m(x) = \sum_{k=1}^{K} \beta_k^m \eta_k(x), \tag{5.10}$$

where m is the harmonic number.

To determine the coefficients of this series it is convenient to use the projection conditions once again, integrating over Ω the length, area or volume of an elastic body

$$\int_\Omega \left(N\eta^m - (m\lambda)^2 R\eta^m - f^m\right)\eta_k \mathrm{d}\Omega = 0. \tag{5.11}$$

Taking into account the self-adjointness of operator N

$$\int_\Omega N\eta^m \eta_k \mathrm{d}\Omega = \int_\Omega \eta^m N\eta_k \mathrm{d}\Omega,$$

where η^m represents the series (5.10) in which index k is replaced by j. Next, we use the equation $N\eta_k - \lambda_k^2 R\eta_k = 0$ for the forms and frequencies of free vibrations as well as the condition of R-orthonormality of the natural forms $\int_\Omega R\eta_j\eta_k \mathrm{d}\Omega = \delta_{jk}$.

As a result, we obtain

$$\left(\lambda_k^2 - (m\lambda)^2\right)\beta_k^m - f_k^m = 0, \text{ where } f_k^m = \int_\Omega f^m \eta_k \, \mathrm{d}\Omega$$

or

$$\beta_k^m = \frac{f_k^m}{\lambda_k^2 - (m\lambda)^2}. \tag{5.12}$$

The greater the number of terms in expansion (5.10) and the number of terms in the harmonic series (5.6), the greater the "resonance" possibility when $\lambda_k^2 - (m\lambda)^2 = 0$ and Eq. (5.9) according to (5.12) does not have a finite solution. An exception is when the corresponding coefficient f_k^m is also equal to zero and the magnitude of $\beta_k^m = \frac{0}{0}$ can take an arbitrary value. We can avoid this situation by assuming the existence of distributed viscous friction. In doing so, Eq. (5.1) can be replaced as follows

$$N\eta + G\dot{\eta} + R\ddot{\eta} - f = 0, \quad f = f(x,t), \quad \eta = \eta(x,t), \tag{5.13}$$

where $G = G(x)$ is the operator of viscous friction to which specific expressions can be accordingly given in the cases of a one-, two- or three-dimensional body. We restrict ourselves to the assumption that the operator $G = G(x)$ is positive definite, and therefore self-adjoint. Representing the solution of polyharmonic oscillations once again as a harmonic series (5.6), using the projection conditions and taking the friction forces into account

$$\frac{1}{T}\int_0^T \mathrm{e}^{-im\lambda t}(N\eta + G\dot{\eta} + R\ddot{\eta} - f)\mathrm{d}t = 0, \tag{5.14}$$

instead of Eq. (5.9), we obtain for each harmonic component of oscillations

$$N\eta^m + im\lambda\, G\,\eta^m - (m\lambda)^2 R\eta^m - f^m = 0. \tag{5.15}$$

Following the spectral method, solution of this equation can again be represented as a series of the forms of free oscillations (5.10). Determination of the coefficients of this series can be brought about using the projection condition

$$\int_\Omega \left(N\eta^m + im\lambda\, G\eta^m - (m\lambda)^2 R\eta^m - f^m\right)\eta_k(x)\mathrm{d}\Omega = 0. \tag{5.16}$$

At

$$G(x) = \gamma R(x), \quad \gamma > 0, \tag{5.17}$$

instead of (5.12), we obtain

$$\beta_k^m = \frac{f_k^m}{\lambda_k^2 - (m\lambda)^2 + im\lambda\gamma}. \tag{5.18}$$

The denominator here does not vanish (i.e. become zero), even in the case of resonance at $\lambda_k^2 - (m\lambda)^2 = 0$. If the assumption-hypothesis (5.17) fails in the general case, a separate formula for β_k^m cannot be obtained and the problem is reduced to the system of equations

$$\left(\lambda_k^2-(m\lambda)^2\right)\beta_k^m+im\lambda\sum_{j=1}^{K}\gamma_{k,j}\beta_j^m-f_k^m=0,$$
$$\gamma_{k,j}=\int\limits_{\Omega}G\eta_k\eta_j\mathrm{d}\Omega,\quad k=1,2,\ldots K. \tag{5.19}$$

When friction is small and retaining only the diagonal elements in the matrix γ_{kj}, we can obtain a formula for determination of the coefficients β_k^m

$$\beta_k^m=\frac{f_k^m}{\lambda_k^2-(m\lambda)^2+im\,\lambda\gamma_k},\quad \gamma_k=\gamma_{kk}=\int\limits_{\Omega}G\eta_k\eta_k d\Omega. \tag{5.20}$$

Physically, this means that friction is significantly taken into account only for resonant terms when the difference $\lambda_k^2-(m\lambda)^2$ is zero or close to it.

5.3 Harmonic Analysis and the Variational Difference Method

In Chap. 4 we used the variational difference method to solve the problem of harmonic vibrations of a straight rod that took viscous friction into account. The results can easily be extended to the case of polyharmonic oscillations. This can be done by placing

$$p(x,t)=\sum_{m=-M}^{M}p^m(x)\mathrm{e}^{im\lambda t},\quad \mathrm{e}(x,t)=\sum_{m=-M}^{M}\mathrm{e}^m(x)\mathrm{e}^{im\lambda t} \tag{5.21}$$

in Eqs. (4.76) instead of (4.78), (4.79).

Accordingly, a periodic solution can be sought similar to (4.80) in the form

$$u(x,t)=\sum_{m=-M}^{M}u^m\mathrm{e}^{im\lambda t},\quad Q(x,t)=\sum_{m=-M}^{M}Q^m\mathrm{e}^{im\lambda t}, \tag{5.22}$$

where $\lambda=\frac{2\pi}{T}, T$ is the period of oscillations; and M is the number of harmonics taken into account.

Equations (4.83) and formulas (4.84)–(4.86) retain the form, but they will also have

$$m=-M,\ldots,M. \tag{5.23}$$

5.4 Periodic Oscillations (Dependence of the Equation Operators on Time)

We assumed above that the operators in Eq. (5.13) depend only on the coordinate x and not on time t. Assuming this in the case of polyharmonic vibrations for each harmonic component, a separate ordinary differential equation of the form (5.9) or (5.15) can be obtained.

Let us now suppose that the coefficients in Eq. (5.13) depend not only on the coordinate x, but also on the periodic functions of time t, $0 \le t \le T$. For definiteness, we accept that the operator of elasticity $B = B(x,t)$ is time dependent in the formula for the operator N having the form $N = DB^{-1}D^*$. We also introduce the inverse operator $C = B^{-1}, C = C(x,t)$. In doing so, only the computation of the following integral will be changed in the projection conditions (5.7) or (5.14)

$$\frac{1}{T}\int\limits_0^T e^{-im\lambda t} N\eta \, dt = \frac{1}{T}\int\limits_0^T e^{-im\lambda t} DC(x,t)D^* \sum_{s=-S}^{S} \eta^{(s)}(x) e^{is\lambda t}$$
$$= \sum_{s=-S}^{S} DC_{ms} D^* \eta^{(s)},$$

where

$$C_{ms} = C_{ms}(x) = \frac{1}{T}\int\limits_0^T C(x,t) e^{i(s-m)\lambda t} dt. \qquad (5.24)$$

Moreover, it is only necessary to compute this integral for $m = 1$. The remaining rows of the matrix (5.24) can be obtained by shifting the previous row leftward by the number of the mth row. Matrices of this type are called “Toeplitz matrixes”. (For example, to calculate such a matrix in the programming language Matlab it is enough to call the function “toeplitz”.) Thus, when $B = B(x,t)$, else $C = C(x,t)$, the problem of forced periodic oscillations of an elastic body with the availability of viscous friction can be reduced to a system of ordinary differential equations, recorded in the operator form

$$\sum_{s=-S}^{S} DC_{ms} D^* \eta^{(s)} + im\lambda G\eta^m - (m\lambda)^2 R\eta^m - f^m = 0. \qquad (5.25)$$

5.5 Nonstationary Oscillations and Harmonic Analysis

Let us replace one second-order Eq. (5.13) by two first-order equations (bearing in mind the time derivative)

$$\begin{aligned} R\dot{\nu} &= -N\eta - G\nu + f, \\ \dot{\eta} &= \nu. \end{aligned} \tag{5.26}$$

Since the operator R is positive definite, it has a bounded inverse R^{-1}, and instead of (5.26) it is possible to write

$$\begin{aligned} \dot{\nu} &= R^{-1}(-N\eta - G\nu + f), \\ \dot{\eta} &= \nu. \end{aligned} \tag{5.27}$$

In principle, the system of Eqs. (5.27) under the initial conditions (5.3) can be solved by integrating the equations of time using known numerical methods, such as the Runge-Kutta method. However, this technique is of limited utility in the case of high-frequency oscillations. This is explained by rapid change in the functions sought; hence, the need to carry out integration with very small increments in time so that significant accumulation of rounding errors can take place. Therefore, the method used to integrate Eqs. (5.26), based on the use of harmonic analysis, can be helpful.

Turning to consideration of this method, we again suppose that the coefficients of Eqs. (5.26) depend only on the coordinate x and that

$$|f(x,t)| \leq F, \quad t \geq 0, \tag{5.28}$$

where F is some positive number.

Let us introduce the functions

$$\begin{aligned} &\psi(x,t) = \eta(x,t)\,\mathrm{e}^{-qt}, \quad \varphi(x,t) = \nu(x,t)\mathrm{e}^{-qt}, \\ &t \geq 0, \quad q > 0, \end{aligned} \tag{5.29}$$

$$\psi(x,0) = \eta(x,0) = \eta_0(x), \quad \varphi(x,0) = \nu(x,0) = \nu_0(x), \tag{5.30}$$

such that

$$\eta(x,t) = \psi(x,t)\mathrm{e}^{qt}, \quad \nu(x,t) = \varphi(x,t)\mathrm{e}^{qt}. \tag{5.31}$$

An approximate solution of Eqs. (5.26) can be found on a sufficiently large but finite interval $0 \leq t \leq T$.

Let us represent the functions $\varphi(x,t)$ and $\varphi(x,t)$ as the harmonic series

$$\begin{aligned}\psi(x,t) &= \sum_{m=-M}^{M} \psi^m(x)e^{im\lambda t},\\ \varphi(x,t) &= \sum_{m=-M}^{M} \varphi^m(x)e^{im\lambda t}, \quad \lambda = \frac{2\pi}{T}.\end{aligned} \tag{5.32}$$

Taking formulas (5.31) into account, the projection conditions for the system (5.26) can be written as follows

$$\begin{aligned}&\frac{1}{T}\int_0^T e^{-(q+im\lambda)t}\left(N(\psi e^{qt}) + G(\varphi e^{qt}) + R\frac{\partial(\varphi e^{qt})}{\partial t} - f\right)dt = 0,\\ &\frac{1}{T}\int_0^T e^{-(q+im\lambda)t}\left(\frac{\partial(\psi e^{qt})}{\partial t} - \varphi e^{qt}\right)dt = 0, \quad m = -M,\ldots,M.\end{aligned} \tag{5.33}$$

Using the rule of integration by parts and the periodicity property of the harmonic functions $e^{im\lambda 0} = e^{im\lambda T} = 1$, it is possible, in particular, to compute the integral involved in the second of these equations

$$\begin{aligned}&\frac{1}{T}\int_0^T e^{-(q+im\lambda)t}\frac{\partial(\psi(x,t)e^{qt})}{\partial t}dt = \frac{1}{T}e^{-(q+im\lambda)t}\psi(x,t)e^{qt}\Big|_{t=0}^{t=T}\\ &+(q+im\lambda)\frac{1}{T}\int_0^T e^{-(q+im\lambda)t}\psi(x,t)e^{qt}dt = \frac{1}{T}(\psi(x,T)-\psi(x,0))\\ &+(q+im\lambda)\frac{1}{T}\int_0^T e^{-im\lambda t}\psi(x,t)dt.\end{aligned}$$

By virtue of (5.28), (5.29) and making viscous friction available by choosing the quantities T and q, it is possible to make the magnitude $|\psi(x,T)|$ negligibly small. The same reasoning can be applied to $|\nu(x,T)|$. Then, in view of the first of the initial conditions (5.30), we get

$$\begin{aligned}&\frac{1}{T}\int_0^T e^{-(q+im\lambda)t}\frac{\partial(\psi(x,t)e^{qt})}{\partial t}dt\\ &= -\frac{1}{T}\eta_0(x) + (q+im\lambda)\psi^m(x).\end{aligned} \tag{5.34}$$

The formula for the coefficients of the first of the series (5.32) is used in (5.34)

$$\psi^m(x) = \frac{1}{T}\int_0^T e^{-im\lambda t}\psi(x,t)\mathrm{d}t.$$

Analogously

$$\begin{aligned}\frac{1}{T}\int_0^T e^{-(q+im\lambda)t}\frac{\partial(\varphi(x,t)e^{qt})}{\partial t}\mathrm{d}t \\ = -\frac{1}{T}v_0(x) + (q+im\lambda)\varphi^m(x), \qquad (5.35)\\ \varphi^m(x) = \frac{1}{T}\int_0^T e^{-im\lambda t}\varphi(x,t)\mathrm{d}t.\end{aligned}$$

In view of these transformations, we obtain the operational equations for each harmonic

$$\begin{aligned} N\psi^m + G\varphi^m + (q+im\lambda)R\varphi^m - \phi^m = 0, \\ (q+im\lambda)\psi^m - \varphi^m - \chi^m = 0, \end{aligned} \qquad (5.36)$$

where $\phi^m = f^m + \frac{1}{T}Rv_0, \chi^m = \frac{1}{T}\eta_0$, and $f^m(x) = \frac{1}{T}\int_0^T e^{-im\lambda t}f(x,t)e^{-qt}\mathrm{d}t$.

Having found $\psi^m = \psi^m(x)$ and $\varphi^m = \varphi^m(x)$ for all harmonics from (5.36), we can approximately determine $\psi(x,t)$ and $\eta(x,t)$ as well as $\varphi(x,t)$ and $v(x,t)$ by formulas (5.31) and (5.32).

Note that the methods outlined for solving the problems of stationary and nonstationary vibrations can be considered as discrete analogues of the methods using the Fourier transform and the Fourier-Laplace transform, respectively [1, 3].

5.6 Oscillations Close to Periodic

It is not uncommon in engineering for mechanical or electromechanical systems to bring about high-frequency vibrations, but the system parameters and therefore the characteristics of the oscillation process are slowly changing in time.

For simplicity of presentation let us consider high-frequency forced linear oscillations of an elementary mechanical system, namely vibrations of a lumped mass attached to a spring. Let us denote by $c(\tau)$ and $f(\tau, t)$, respectively, the spring stiffness coefficient and the external force acting on the mass divided by the quantity of this mass. We denote by $\gamma(\tau)$ the coefficient of friction acting on the mass, also divided by the mass value. Let us assume that the indicated quantities depend on the slow time τ. Moreover, let us suppose that at the fixed value of τ the external force $f(\tau, t)$ is a periodic function of fast time t with the frequency $\omega(\tau)$ and the period $T(\tau) = \frac{2\pi}{\omega(\tau)}$. The implication is that there exist periodic oscillations with the same frequency and period. Such movements are close to those in the following discussion. Next, let us consider that the slow time τ dependence on the fast time t is preset

$$\frac{d\tau}{dt} = \varepsilon(t), \quad \tau(0) = 0, \quad |\varepsilon(t)| \leq \varepsilon_{max}, \quad 0 \leq t \leq t_{max} < \infty, \tag{5.37}$$

and ε_{max} is the small number.

The notation introduced allows the equation for displacements of vibrations of the lumped mass attached to the spring to take the form

$$\frac{d^2u}{dt^2} + \gamma(\tau)\frac{du}{dt} + c(\tau)u = f(\tau, t). \tag{5.38}$$

Since we are dealing with high-frequency oscillations, it is fair to assume that at τ of (5.37) and a limited range of variation of the fast time t

$$|\omega(\tau)| \gg 0. \tag{5.39}$$

The equation of oscillations (5.38) can be replaced with a system of two equations of the first order

$$\begin{aligned} &\frac{dv}{dt} + \gamma(\tau)v + c(\tau)u - f(\tau, t) = 0, \\ &\frac{du}{dt} - v = 0. \end{aligned} \tag{5.40}$$

Let us introduce a new variable $\theta = \theta(t)$, the derivative of which with respect to time t is equal to the instantaneous frequency of the external periodic force

$$\frac{d\theta}{dt} = \omega(\tau), \tag{5.41}$$

with $\theta = 0$ at $t = 0$.

Obviously

$$\frac{d}{dt} = \frac{d}{d\theta}\frac{d\theta}{dt} = \frac{d}{d\theta}\omega. \tag{5.42}$$

Formula (5.41) assigns its own value of θ to each value of t.

Taking formula (5.42) into account, Eqs. (5.40) can be rewritten in such a way as

$$\begin{aligned}&\frac{\mathrm{d}v}{\mathrm{d}\theta}\omega(\tau)+\gamma(\tau)v+c(\tau)u-f(\tau,\theta)=0,\\&\frac{\mathrm{d}u}{\mathrm{d}\theta}\omega(\tau)-v=0.\end{aligned}\tag{5.43}$$

For the fixed value of slow time τ in which $0\leq t\leq T(\tau)$, we also get $0\leq\theta(t)\leq 2\pi$. This means that if $\omega=const$, then according to (5.41) $\theta=\omega t$ and, when $t=T$, we also have $\theta=\omega T=\frac{2\pi}{T}T=2\pi$. As this takes place the nonhomogeneous term in Eqs. (5.40), being a periodic function of time t, where $0\leq t\leq T(\tau)$, is also a periodic function of θ in Eqs. (5.43), where $0\leq\theta\leq 2\pi$. Therefore, an approximate solution of the equation system (5.43) can be represented as a series

$$u(\tau,\theta)=\sum_{m=-M}^{M}u^m(\tau)\mathrm{e}^{im\theta},\quad \nu(\tau,\theta)=\sum_{m=-M}^{M}\nu^m(\tau)\mathrm{e}^{im\theta}.\tag{5.44}$$

Hence, $\tau=\tau(t),\theta=\theta(t)$, which defines the parametric dependence

$$\tau=\tau(\theta).\tag{5.45}$$

Note that the functions $\mathrm{e}^{im\theta}$ have the property of orthonormality

$$\frac{1}{2\pi}\int_0^{2\pi}\mathrm{e}^{im\theta}\mathrm{e}^{-is\theta}\mathrm{d}\theta=\delta_{ms},\quad m,s=-M,\ldots,M,\tag{5.46}$$

where δ_{ms} is the Kronecker symbol. We can write the projection conditions as

$$\begin{aligned}&\frac{1}{2\pi}\int_0^{2\pi}\mathrm{e}^{-is\theta}\left(\frac{\mathrm{d}\nu}{\mathrm{d}\theta}\omega+\gamma\nu+cu-f\right)\mathrm{d}\theta=0,\\&\frac{1}{2\pi}\int_0^{2\pi}\mathrm{e}^{-is\theta}\left(\frac{\mathrm{d}u}{\mathrm{d}\theta}\omega-\nu\right)\mathrm{d}\theta=0,\quad s=-M,\ldots,M.\end{aligned}\tag{5.47}$$

We represent $\nu(\tau,\theta)$ and $u(\tau,\theta)$ in the form of series (5.44) and take into account that

$$\begin{aligned}&\frac{\mathrm{d}\nu}{\mathrm{d}\theta}=\sum_{m=-M}^{M}\left(\frac{\mathrm{d}\nu^m}{\mathrm{d}\theta}+im\nu^m\right)\mathrm{e}^{im\theta},\\&\frac{\mathrm{d}u}{\mathrm{d}\theta}=\sum_{m=-M}^{M}\left(\frac{\mathrm{d}u^m}{\mathrm{d}\theta}+imu^m\right)\mathrm{e}^{im\theta}.\end{aligned}\tag{5.48}$$

Then, it becomes

$$\frac{1}{2\pi}\sum_{m=-M}^{M}\int_0^{2\pi}\frac{\mathrm{d}v^m}{\mathrm{d}\theta}\omega\,\mathrm{e}^{i(m-s)\theta}\mathrm{d}\theta=\frac{1}{2\pi}\sum_{m=-M}^{M}\int_0^{2\pi}(-(im\omega+\gamma)v^m-cu^m)\mathrm{e}^{i(m-s)\theta}\mathrm{d}\theta$$
$$+\frac{1}{2\pi}\int_0^{2\pi}f\mathrm{e}^{-is\theta}\mathrm{d}\theta,$$

$$\frac{1}{2\pi}\sum_{m=-M}^{M}\int_0^{2\pi}\frac{\mathrm{d}u^m}{\mathrm{d}\theta}\omega\,\mathrm{e}^{i(m-s)\theta}\mathrm{d}\theta=\frac{1}{2\pi}\sum_{m=-M}^{M}\int_0^{2\pi}(v^m-im\omega u^m)\mathrm{e}^{i(m-s)\theta}\mathrm{d}\theta.$$

For computation of the integrals we consider slow time τ as a parameter and integrate only over θ. Taking the condition of orthonormality (5.46) into account, we finally arrive at the system of equations for the coefficients of the series (5.44)

$$\begin{aligned}\frac{\mathrm{d}v^m(\tau)}{\mathrm{d}\theta}&=-imv^m(\tau)+\frac{1}{\omega(\tau)}(-\gamma(\tau)v^m(\tau)-c(\tau)u^m(\tau)+f^m(\tau)),\\ \frac{\mathrm{d}u^m(\tau)}{\mathrm{d}\theta}&=-imu^m(\tau)+\frac{1}{\omega(\tau)}v^m(\tau),\end{aligned}\tag{5.49}$$

where $f^m(\tau)=\frac{1}{2\pi}\int_0^{2\pi}f(\tau,\theta)\mathrm{e}^{-im\theta}\mathrm{d}\theta$.

Let us now suppose that motion with slowly varying parameters, being close to periodic, starts at $\tau=0$ as a purely periodic motion. Representing the initial periodic motion in the form of an approximate Fourier series expansion in the powers $\mathrm{e}^{im\theta}$, $m=-M,\ldots,M$, and determining the coefficients of this series, we obtain the initial values $u^m(0)$ and $v^m(0)$ that should be added to the system of Eqs. (5.49).

The solution of Eqs. (5.49) can be found by numerical integration (e.g. by the Runge-Kutta method). It is important that all the coefficients on the right-hand sides of Eqs. (5.49) are functions of the slow time τ. Therefore, it is not necessary here to carry out integration with a small step, as would be required at integration of the initial Eqs. (5.40) with respect to the oscillating functions $u(\tau,t),\nu(\tau,t)$.

Formulas (5.44) and (5.49) allow the approximate solution of the initial problem in dependence on θ to be obtained. Having constructed the function $\theta(t)$ according to (5.41) and knowing $\tau(t)$ by virtue of (5.37), it is an easy matter to calculate the values of $u(t)=u(\tau,t)$ and $\nu(t)=v(\tau,t)$ at any moment of time t, $0\leq t\leq{}_{\max}<\infty$.

The method proposed can be extended to the case of a mechanical system with a finite number of degrees of freedom. In this the required functions of time $u(t)$, $\nu(t)$, as well as the coefficients of the series (5.44) turn into vector functions.

As a first approximation, by maintaining only one harmonic in the expansions (5.44), the method outlined for solving the problem of vibrations close to periodic essentially coincides with the van der Pol method [6]. Justification of the method described can be found in [4].

References

1. Iofin, L. E., & Fridman, V. M. (1979). Generalization of method of dynamic stiffness to the case of non-stationary oscillations of complex elastic systems. In *Proceedings of the U.S.S.R. Academy of Sciences, Mechanics of Solids*, (3, pp. 158–163) (in Russian).
2. Kutta, W. (1901). Beitrag zur naherungsweisen Integration totaler Differentialgleichungen. *Zeitschrift für Mathematik und Physik, 46,* 435–453.
3. Marchuk, G. I. (1982). *Methods of Numerical Mathematics*. (p. 510), Springer.
4. Merkin, M. R., & Fridman, V. M. (1981). Projection method for solving the problem of forced non-stationary oscillations in nonlinear systems with slowly varying parameters. *Academy of Sciences of the U.S.S.R., Applied Mathematics and Mechanics*, *45*(1), 71–79 (in Russian).
5. Runge, C. (1895). Über die numerische Ausflösung von Differentialgleichungen. *Mathematische Annales, 46,* 167–178.
6. Van der Pol, B. (1926). On relaxation-oscillation. *London, Edinburgh and Dublin Philosophical Magazine and Journal of Science*, *2*(7), 978–992.

Chapter 6
Discontinuous Functions. Complicated Boundary Conditions

6.1 Longitudinal Vibrations of a Rectilinear Rod

The problems of oscillations of elastic bodies with simple homogeneous boundary conditions have been considered above. However, let is abandon this assumption and consider that boundary conditions can be complex (mixed) homogeneous and nonhomogeneous as well. Let us start for simplicity and obviousness with the simplest problem: the harmonic longitudinal vibrations of a rectilinear rod. The equations of the amplitude values of tensile force $Q = Q(x)$ and displacement $u = u(x)$ in this case have the form (1.14)

$$\begin{aligned} -\frac{\mathrm{d}Q}{\mathrm{d}x} - \lambda^2 \rho u - p = 0, \\ \frac{\mathrm{d}u}{\mathrm{d}x} - bQ - e = 0, \end{aligned} \tag{6.1}$$

where, as before, x is the coordinate measured along the rod, $0 \le x \le l$; l is the rod length; $p = p(x)$ and $e = e(x)$ are the amplitudes of distributed external load and deformation from external causes; $\rho = \rho(x)$ and $b = b(x)$ are the characteristics of distributed inertia and elasticity; and λ is the frequency of harmonic vibrations.

Let us now examine the case when point forces operate and lumped masses are fixed on a rod, along with the distributed loads, deformations and distributed external actions on it. Moreover, assume that the rod can be divided into a finite number of parts that are connected with the lumped springs, and these springs are subjected to concentrated deformations from external causes.

For the coordinate counted along the rod, we need another notation s, $0 \le s \le l$, and a function called the "delta function" $\delta(x - s)$ such that the following formula holds

V. Fridman, *Theory of Elastic Oscillations*, Foundations of Engineering Mechanics, DOI 10.1007/978-981-10-4786-2_6

$$\varphi(s) = \int_0^l \varphi(x)\delta(x-s)\mathrm{d}x, \tag{6.2}$$

as long as the function $\varphi(x)$ is continuous at the point $x = s$.

Let us suppose that external point forces $p_k, k = 1, 2, \ldots, K$, operate in the cross-sections of the bar with coordinates s_k and, in accordance with this system of forces, set the distributed load as

$$\widehat{p}(x) = \sum_{k=1}^{K} p_k\delta(x - s_k). \tag{6.3}$$

Let us calculate the work of load (6.3) on the displacement $u = u(x)$, assuming that this displacement is continuous at the points $x = s_k$. Using formula (6.2) we get

$$\begin{aligned}\Pi &= \frac{1}{2}\int_0^l \widehat{p}(x)u(x)\mathrm{d}x = \frac{1}{2}\int_0^l \sum_{k=1}^{K} p_k\delta(x - s_k)u(x)\mathrm{d}x \\ &= \frac{1}{2}\sum_{k=1}^{K} p_k u(s_k).\end{aligned} \tag{6.4}$$

It can be seen that formally replacing the abovementioned system of point forces with the distributed load (6.3), as a result of calculating the integral (6.4) and taking the properties of the delta function (6.2) into account, we automatically obtain a formula for calculation of the work of these point forces.

Should the distributed load denoted by $\tilde{p}(x)$ occur in addition to the point forces, then it is possible to compose the total load as

$$p(x) = \tilde{p}(x) + \widehat{p}(x), \tag{6.5}$$

which would include both the distributed load and concentrated forces. In like manner let us proceed to considering the concentrated strain and assume the presence of the lumped inertial and elastic characteristics of the rod. Namely, we assume that

$$\begin{aligned} &e(x) = \tilde{e}(x) + \widehat{e}(x),\ \rho(x) = \tilde{\rho}(x) + \widehat{\rho}(x), \\ &b(x) = \tilde{b}(x) + \widehat{b}(x), \end{aligned} \tag{6.6}$$

where $\tilde{e}(x), \tilde{\rho}(x), \tilde{b}(x)$ are the distributed characteristics of strain, inertia and elasticity. Moreover

$$\widehat{e}(x) = \sum_{j=1}^{J} e_j \delta(x - s_j), \quad \widehat{\rho}(x) = \sum_{k=1}^{K} \rho_k \delta(x - s_k),$$
$$\widehat{b}(x) = \sum_{j=1}^{J} b_j \delta(x - s_j). \tag{6.7}$$

Taking formulas (6.3), (6.5)–(6.7) into account, the system of Eq. (6.1) can be rewritten as follows

$$-\frac{dQ(x)}{dx} - \lambda^2 \left(\tilde{\rho}(x) + \sum_{k=1}^{K} \rho_k \delta(x - s_k) \right) u(x) - \left(\tilde{p}(x) + \sum_{k=1}^{K} p_k \delta(x - s_k) \right) = 0,$$
$$\frac{du(x)}{dx} - \left(\tilde{b}(x) + \sum_{j=1}^{J} b_j \delta(x - s_j) \right) Q(x) - \left(\tilde{e}(x) + \sum_{j=1}^{J} e_j \delta(x - s_j) \right) = 0. \tag{6.8}$$

Stretching (contracting) the longitudinal force $Q(x)$ of the rod has discontinuities at the points of application of the concentrated forces. That is

$$Q(x+0) - Q(x-0) = -(\lambda^2 \rho_k u(x) + p_k), \text{ where } \quad x = s_k \tag{6.9}$$

as long as the function $u(x)$ is continuous at these points. Similarly, the concentrated strain from external factors e_j and the concentrated elasticity b_j are the cause of discontinuities of the displacement $u(x)$ at the points $x = s_j$

$$u(x+0) - u(x-0) = b_j Q(x) + e_j, \tag{6.10}$$

hence, the function $Q(x)$ should be continuous at the same points.

The presence of jumps of stresses or displacements where concentrated forces are applied or where concentrated strain is placed is understandable from physical considerations. However, equalities (6.9), (6.10) can be obtained formally from Eq. (6.8). This means that in formula (6.2) we can put

$$\varphi(x) = 0, \; 0 \le x < s - 0.5\varepsilon,$$
$$\varphi(x) = 1, \; s - 0.5\varepsilon \le x \le s + 0.5\varepsilon, \tag{6.11}$$
$$\varphi(x) = 0, \; s + 0.5\varepsilon < x \le l,$$

where $\varepsilon > 0$. We obtain

$$1 = \int_{s-0.5\varepsilon}^{s+0.5\varepsilon} \delta(x - s) dx, \tag{6.12}$$

since the function $\varphi(x)$ is continuous at the point $x = s$ and $\varphi(s) = 1$.

In formula (6.12) the positive number ε can be arbitrarily small, such that when $\varepsilon \to 0$

$$\lim_{\varepsilon\to 0}\left(\int_{s-0.5\varepsilon}^{s+0.5\varepsilon} \delta(x-s)\mathrm{d}x\right) = \int_{s-0}^{s+0} \delta(x-s)\mathrm{d}x$$

and therefore

$$\int_{s-0}^{s+0} \delta(x-s)\mathrm{d}x = 1. \tag{6.13}$$

Moreover, $\int_{s-0.5\varepsilon}^{s+0.5\varepsilon} \frac{\mathrm{d}Q}{\mathrm{d}x}\mathrm{d}x = Q(s+0.5\varepsilon) - Q(s-0.5\varepsilon)$ regardless of whether the function $Q(x)$ is continuous or has discontinuities limited in magnitude on the interval of integration. Therefore, in the limit at $\varepsilon \to 0$ we can write

$$\int_{s-0}^{s+0} \frac{\mathrm{d}Q}{\mathrm{d}x}\mathrm{d}x = Q(s+0) - Q(s-0). \tag{6.14}$$

Also

$$\int_{s-0}^{s+0} \frac{\mathrm{d}u}{\mathrm{d}x}\mathrm{d}x = u(s+0) - u(s-0). \tag{6.15}$$

Having applied formulas (6.13)–(6.15) to the first and second Eq. (6.8) at $s = s_k$ and $s = s_j$, respectively, and taking into account that

$$\int_{s-0}^{s+0} \tilde{p}(x)\mathrm{d}x = 0, \quad \int_{s-0}^{s+0} \tilde{\rho}(x)u(x)\mathrm{d}x = 0,$$

$$\int_{s-0}^{s+0} \tilde{b}(x)Q(x)\mathrm{d}x = 0, \quad \int_{s-0}^{s+0} \tilde{e}(x)\mathrm{d}x = 0,$$

we get to formulas (6.9), (6.10).

Thus, the jump conditions obtained for stress and displacement can be used to compile the boundary conditions at the ends of the rod. Assume for concreteness that on the left end of the bar at $x = 0$ there is a spring with elasticity b_0, and the amplitude of deformation of the spring from external causes is equal to e_0. Use a boundary condition of the type (6.10) and assume that the spring rests on an absolutely rigid wall (i.e. $u(-0) = 0$). Then the boundary condition (6.10) at $x = 0$ takes the form

$$u(+0) = b_0 Q(0) + e_0. \tag{6.16}$$

Similarly, consider that on the right end of the rod at $x = l$ a condition of the type (6.9) is fulfilled, where $Q(l+0) = 0$ (there is a void behind the cross-section of the rod with the coordinate $x = l$). Then the boundary condition transforms into

$$-Q(l-0) = -\left(\lambda^2 \rho_l u(l) + p_l\right), \tag{6.17}$$

where ρ_l and p_l are the lumped mass and point force at $x = l$.

Since the points $(+0)$ and $(l-0)$ belong to the rod, we can simply write

$$u(+0) = u(0), \; Q(l-0) = Q(l). \tag{6.18}$$

Naturally, at $b_0 = 0, e_0 = 0, \rho_l = 0, p_l = 0$ the complex (mixed) nonhomogeneous boundary conditions (6.16)–(6.18) transform into the simple homogeneous conditions

$$u(0) = 0 \text{ and } Q(l) = 0. \tag{6.19}$$

6.2 The Spectral Method Using the Simple Homogeneous Boundary Conditions

It is not infrequent in applied problems when the difficulty of solving the problem of vibrations of an elastic body stems from the complexity of boundary conditions. In particular, it can be difficult to find the forms and frequencies of free oscillations of an elastic body. At the same time, it is possible to specify simple homogeneous boundary conditions that facilitate solution of the spectral problem. For descriptive reasons let us present the simplest example of longitudinal vibrations of a rod with distributed parameters $\rho(x)$, $b(x)$ and distributed load $p(x)$ in the absence of deformation from external causes when $e(x) = 0$. Let us suppose that on the end of the rod at $x = l$ the lumped mass m is fixed and the end $x = 0$ is clamped.

The system of Eqs. (6.8) and the boundary conditions (6.16)–(6.18) in this case take the form

$$
\begin{aligned}
&-\frac{\mathrm{d}Q}{\mathrm{d}x} - \lambda^2 \rho u - p = 0, \\
&\frac{\mathrm{d}u}{\mathrm{d}x} - bQ = 0, \\
&u(0) = 0, \quad Q(l) = \lambda^2 m u(l),
\end{aligned} \tag{6.20}
$$

where the second boundary condition is mixed.

From Eqs. (6.20) it follows that

$$Q = \left(\frac{1}{b}\frac{\mathrm{d}u}{\mathrm{d}x}\right)$$

and

$$-\frac{\mathrm{d}}{\mathrm{d}x}\left(\frac{1}{b}\frac{\mathrm{d}u}{\mathrm{d}x}\right) - \lambda^2 \rho u - p = 0. \tag{6.21}$$

Let us represent the desired solution as a sum of two terms

$$u(x) = \tilde{u}(x) + \widehat{u}(x), \tag{6.22}$$

having extracted the displacement part $\tilde{u}(x)$ satisfying the simple homogeneous conditions, we get

$$\tilde{u}(0) = 0, \quad \tilde{Q}(l) = \frac{1}{b(l)}\frac{\mathrm{d}\tilde{u}}{\mathrm{d}x}(l) = 0. \tag{6.23}$$

Let us suppose we have determined the forms and frequencies of the free longitudinal vibrations of a rod that meet these simple homogeneous boundary conditions

$$\tilde{u}_i(0) = 0, \quad \tilde{Q}_i(l) = \frac{1}{b(l)}\frac{\mathrm{d}\tilde{u}_i}{\mathrm{d}x}(l) = 0 \tag{6.24}$$

and the homogeneous equation

$$-\frac{\mathrm{d}}{\mathrm{d}x}\left(\frac{1}{b}\frac{\mathrm{d}\tilde{u}_i}{\mathrm{d}x}\right) - (\tilde{\lambda}_i)^2 \rho \tilde{u}_i = 0. \tag{6.25}$$

Let us represent the function $\tilde{u}(x)$ of formula (6.22) as a series expansion in the forms of free oscillations $\tilde{u}_i(x), i = 1, 2, \ldots, I$

$$\tilde{u}(x) = \sum_{i=1}^{I} \tilde{\alpha}_i \tilde{u}_i(x). \tag{6.26}$$

The term $\widehat{u}(x)$ in Eq. (6.22) has been introduced to fulfill the second boundary condition (6.20). All that is needed to do this is determine one form of the free oscillations of the rod, keep the same boundary condition at $x = 0$ as the one in conditions (6.24) and accept the conjugate boundary condition at $x = l$

$$-\frac{\mathrm{d}}{\mathrm{d}x}\left(\frac{1}{b}\frac{\mathrm{d}\widehat{u}_1}{\mathrm{d}x}\right) - (\widehat{\lambda}_1)^2 \rho \widehat{u}_1 = 0, \quad \widehat{u}_1(0) = 0, \widehat{u}_1(l) = 0. \tag{6.27}$$

(we replace $\tilde{Q}(l) = 0$ in conditions (6.24) by $\widehat{u}(l) = 0$).

In accordance with (6.24), note that $\tilde{Q}_i(l) = 0$ at $x = l$ and, generally speaking, $\tilde{u}_i(l) \neq 0$, and the converse is true for the conjugate boundary condition (6.27) with $\widehat{u}_1(l) = 0$ and possibly $\widehat{Q}_1(l) = \frac{1}{b(l)}\frac{\mathrm{d}\widehat{u}_1}{\mathrm{d}x}(l) \neq 0$. Let

$$\widehat{u}(x) = \widehat{\beta}_1 \widehat{u}_1(x). \tag{6.28}$$

Taking formulas (6.22), (6.27), (6.28) into account, we can rewrite Eq. (6.21) as follows

$$-\frac{\mathrm{d}}{\mathrm{d}x}\left(\frac{1}{b}\frac{\mathrm{d}\tilde{u}}{\mathrm{d}x}\right) - \lambda^2 \rho \tilde{u} - \tilde{p} = 0, \tag{6.29}$$

where there will be

$$\tilde{p} = p + (\lambda^2 - (\widehat{\lambda}_1)^2)\rho \widehat{\beta}_1 \widehat{u}_1 \tag{6.30}$$

Let us write the projection conditions for Eq. (6.29)

$$\int_0^l \left(-\frac{\mathrm{d}}{\mathrm{d}x}\left(\frac{1}{b}\frac{\mathrm{d}\tilde{u}}{\mathrm{d}x}\right) - \lambda^2 \rho \tilde{u} - \tilde{p}\right) \tilde{u}_i(x)\mathrm{d}x = 0, \quad i = 1, 2, \ldots, I. \tag{6.31}$$

Substitute hither $\tilde{u}(x)$ from the formula (6.26), replacing previously in this formula the index i by the index j and taking the corresponding integrals by parts, also taking into account that, according to Eqs. (6.25) and (6.27), there will be

$$-\frac{\mathrm{d}}{\mathrm{d}x}\left(\frac{1}{b}\frac{\mathrm{d}\tilde{u}_j}{\mathrm{d}x}\right) = (\tilde{\lambda}_j)^2 \rho \tilde{u}_j,$$

$$-\frac{\mathrm{d}}{\mathrm{d}x}\left(\frac{1}{b}\frac{\mathrm{d}\widehat{u}_1}{\mathrm{d}x}\right)=(\widehat{\lambda}_1)^2\rho\widehat{u}_1,$$

and using the property of ρ-orthonormality of the natural forms $\tilde{u}_i(x)$ we get

$$\int_0^l \rho\tilde{u}_i\tilde{u}_j\mathrm{d}x=\delta_{ij},$$

where, as before, δ_{ij} is the Kronecker symbol. Having also taken formula (6.30) into account, we obtain

$$-\left(\lambda^2-(\tilde{\lambda}_i)^2\right)\tilde{\alpha}_i-(\lambda^2-(\widehat{\lambda}_1)^2)\rho_{1i}\widehat{\beta}_1-p_i=0, \tag{6.32}$$

where $\rho_{1i}=\int_0^l \rho\widehat{u}_1\tilde{u}_i\mathrm{d}x, \quad p_i=\int_0^l p\tilde{u}_i\mathrm{d}x.$

Furthermore, the first boundary condition (6.20) (i.e. the condition $u(0)=0$) is satisfied by virtue of formulas (6.22) (6.24) (6.26) and (6.27). In the left-hand side of the second boundary condition (6.20) there will be

$$Q(l)=\tilde{Q}(l)+\widehat{Q}(l)=\widehat{Q}(l)=\widehat{\beta}_1\frac{1}{b(l)}\frac{\mathrm{d}\widehat{u}_1}{\mathrm{d}x}(l). \tag{6.33}$$

In the right-hand side of the second condition (6.20) the quantity $u(l)$ should be represented in the form of the series (6.26) at $x=l$. As a result, we come to such an equality as

$$\widehat{\beta}_1\frac{1}{b(l)}\frac{\mathrm{d}\widehat{u}_1}{\mathrm{d}x}(l)=\lambda^2 m\sum_{i=1}^{I}\tilde{\alpha}_i\tilde{u}_i(l). \tag{6.34}$$

Substituting $\tilde{\alpha}_i$ in formula (6.32), we can define the value of the coefficient $\widehat{\beta}_1$. Conversely, having $\widehat{\beta}_1$ in the same formula (6.32), we can find the values of the coefficients $\tilde{\alpha}_i, i=1,\ldots,I$. Therefore, using formulas (6.22), (6.26) and (6.28), we can obtain the solution of the initial problem (6.20).

Let us consider another very simple illustrative example: the problem of harmonic bending vibrations of a rectilinear bar, supported at its ends on the springs. The designations of given and unknown quantities are exactly the same as in Chap. 1. From formulas (1.50) and (1.51) at $k=0$ it is not difficult to derive the equation of harmonic oscillations for this displacement problem

$$\frac{\mathrm{d}^2}{\mathrm{d}x^2}\left(\frac{1}{\beta}\frac{\mathrm{d}^2u}{\mathrm{d}x^2}\right)-\lambda^2\rho u-p=0, \tag{6.35}$$

therefore, the bending moment

$$M(x) = -\frac{1}{\beta}\frac{\mathrm{d}^2 u(x)}{\mathrm{d}x^2}. \tag{6.36}$$

Let us write the boundary conditions as

$$\begin{aligned} u(0) &= b_0 Q(0), u(l) = -b_l Q(l), \\ M(0) &= 0, M(l) = 0, \end{aligned} \tag{6.37}$$

where b_0, b_l are the spring elasticity coefficients.

The desired solution can be represented as

$$u(x) = \tilde{u}(x) + \widehat{u}(x), \tilde{M}(x) = -\frac{1}{\beta}\frac{\mathrm{d}^2\tilde{u}(x)}{\mathrm{d}x^2}, \ \widehat{M}(x) = -\frac{1}{\beta}\frac{\mathrm{d}^2\widehat{u}(x)}{\mathrm{d}x^2}, \tag{6.38}$$

where the summand $\tilde{u}(x)$, representing deflection of the bar with absolutely rigid supports on the ends, is extracted. This term in formula (6.38) instead of the mixed boundary conditions (6.37) satisfies the simple homogeneous conditions

$$\tilde{u}(0) = 0, \tilde{u}(l) = 0, \quad \tilde{M}(0) = 0, \tilde{M}(l) = 0. \tag{6.39}$$

The second term in formula (6.38) can be represented in the form

$$\widehat{u}(x) = \widehat{\beta}_1 \widehat{u}_1(x) + \widehat{\beta}_2 \widehat{u}_2(x), \tag{6.40}$$

where the functions $\widehat{u}_j(x), \quad j = 1, 2$, satisfy the homogeneous condition

$$\widehat{M}_j = -\frac{1}{\beta}\frac{\mathrm{d}^2\widehat{u}_j}{\mathrm{d}x^2} = 0, \quad j = 1, 2. \tag{6.41}$$

Namely, let us assume

$$\widehat{u}_1(x) = \frac{x}{l}, \quad \widehat{u}_2(x) = 1 - \frac{x}{l}. \tag{6.42}$$

Substituting $u(x)$ from formula (6.38) into Eq. (6.35), we obtain as a result of (6.41)

$$\frac{\mathrm{d}^2}{\mathrm{d}x^2}\left(\frac{1}{\beta}\frac{\mathrm{d}^2\tilde{u}}{\mathrm{d}x^2}\right) - \lambda^2\rho\tilde{u} - \tilde{p} = 0, \quad \tilde{p} = p + \lambda^2\rho\widehat{u}. \tag{6.43}$$

Let us now represent the function $\tilde{u}(x)$ as a series expansion in the forms of free vibrations of a rod resting on absolutely rigid supports at its ends

$$\tilde{u}(x) = \sum_{i=1}^{I} \tilde{\alpha}_i \tilde{u}_i(x), \tag{6.44}$$

where the forms of free oscillations satisfy the homogeneous equations

$$\frac{\mathrm{d}^2}{\mathrm{d}x^2}\left(\frac{1}{\beta}\frac{\mathrm{d}^2\tilde{u}_i}{\mathrm{d}x^2}\right) - (\tilde{\lambda}_i)^2 \rho \tilde{u}_i = 0 \tag{6.45}$$

and simple homogeneous boundary conditions

$$\tilde{u}_i(0) = 0, \tilde{u}_i(l) = 0, \quad \tilde{M}_i(0) = 0, \tilde{M}_i(l) = 0, \tag{6.46}$$

where

$$\tilde{M}_i(x) = -\frac{1}{\beta}\frac{\mathrm{d}^2\tilde{u}_i(x)}{\mathrm{d}x^2}. \tag{6.47}$$

To determine the coefficients $\tilde{\alpha}_i$ let us compose the projection conditions

$$\int_0^l \left(\frac{\mathrm{d}^2}{\mathrm{d}x^2}\left(\frac{1}{\beta}\frac{\mathrm{d}^2\tilde{u}}{\mathrm{d}x^2}\right) - \lambda^2 \rho \tilde{u} - \tilde{p}\right)\tilde{u}_i \mathrm{d}x = 0 \tag{6.48}$$

or as a result of taking the boundary conditions (6.39), (6.46) into account as well as (6.40), (6.43)–(6.45)

$$-\left(\lambda^2 - (\tilde{\lambda}_i)^2\right)\tilde{\alpha}_i - \lambda^2(\rho_{1i}\widehat{\beta}_1 + \rho_{2i}\widehat{\beta}_2) - p_i = 0, \tag{6.49}$$

where

$$\rho_{ji} = \int_0^l \rho \widehat{u}_j \tilde{u}_i \mathrm{d}x, \; p_i = \int_0^l p \tilde{u}_i \mathrm{d}x, \quad i = 1, 2, \ldots, I, \quad j = 1, 2, \tag{6.50}$$

such that

$$\tilde{\alpha}_i = -\frac{\lambda^2(\rho_{1i}\widehat{\beta}_1 + \rho_{2i}\widehat{\beta}_2) + p_i}{\left(\lambda^2 - (\tilde{\lambda}_i)^2\right)}. \tag{6.51}$$

At the same time, from the first Eq. (1.50) for the case of harmonic oscillations it follows that

$$-\frac{\mathrm{d}Q}{\mathrm{d}x} - \lambda^2 \rho u - p = 0.$$

and, therefore

$$\int_0^l \left(\frac{\mathrm{d}Q}{\mathrm{d}x} + \lambda^2 \rho u + p\right) \widehat{u}_j(x)\mathrm{d}x = 0, \quad j = 1, 2.$$

Using the formula for integration by parts and taking formulas (1.50), (6.37) and (6.42) into account, we get

$$\int_0^l \frac{\mathrm{d}Q}{\mathrm{d}x}\widehat{u}_1(x)\mathrm{d}x = Q(x)\widehat{u}_1(x)|_{x=0}^{x=l} - \int_0^l Q(x)\frac{\mathrm{d}\widehat{u}_1(x)}{\mathrm{d}x}\mathrm{d}x$$

$$= Q(l) - \frac{1}{l}\int_0^l \frac{\mathrm{d}M}{\mathrm{d}x}\mathrm{d}x = Q(l) - \frac{1}{l}(M(l) - M(0)) = Q(l)$$

and, analogously

$$\int_0^l \frac{\mathrm{d}Q}{\mathrm{d}x}\widehat{u}_2(x)\mathrm{d}x = -Q(0).$$

As a result, it turns out that

$$\begin{aligned} Q(0) &= \int_0^l \left(\lambda^2 \rho(\tilde{u} + \widehat{u}) + p\right)\widehat{u}_2(x)\mathrm{d}x, \\ Q(l) &= -\int_0^l \left(\lambda^2 \rho(\tilde{u} + \widehat{u}) + p\right)\widehat{u}_1(x)\mathrm{d}x. \end{aligned} \tag{6.52}$$

This means the first mixed boundary condition (6.37) can be written as follows

$$u(0) = \tilde{u}(0) + \widehat{u}(0) = \widehat{u}(0) = \widehat{\beta}_2$$
$$= b_0 \int_0^l \left(\lambda^2 \rho(\tilde{u} + \widehat{u}) + p\right)\widehat{u}_2(x)\mathrm{d}x.$$

By substituting $\widehat{u} = \widehat{u}(x)$ and $\tilde{u} = \tilde{u}(x)$ for the sums (6.40), (6.44) and interchanging the order of integration and summation, we get

$$\widehat{\beta}_2 = b_0\left(\lambda^2\left(\sum_{i=1}^{I}\rho_{2i}\tilde{\alpha}_i + \widehat{\rho}_{21}\widehat{\beta}_1 + \widehat{\rho}_{22}\widehat{\beta}_2\right) + \widehat{p}_2\right), \tag{6.53}$$

and

$$\widehat{\beta}_1 = b_l\left(\lambda^2\left(\sum_{i=1}^{I}\rho_{1i}\tilde{\alpha}_i + \widehat{\rho}_{11}\widehat{\beta}_1 + \widehat{\rho}_{12}\widehat{\beta}_2\right) + \widehat{p}_1\right), \tag{6.54}$$

where the coefficients ρ_{ji} are given in formula (6.50) and

$$\widehat{\rho}_{ij} = \int_0^l \rho\widehat{u}_i\widehat{u}_j\mathrm{d}x, \quad i,j = 1,2, \quad \widehat{p}_j = \int_0^l p\widehat{u}_j\mathrm{d}x, \quad j = 1,2. \tag{6.55}$$

Having substituted $\tilde{\alpha}_i, \quad i = 1,2,\ldots,I$, into Eqs. (6.53) and (6.54) from formula (6.51), we obtain the system of two equations for the coefficients $\widehat{\beta}_j, j = 1,2$. Using the same formula (6.51) to determine this, we can compute the coefficients $\tilde{\alpha}_i, \quad i = 1,2,\ldots,I$. Using formulas (6.38), (6.40), (6.44), we first find $\widehat{u}(x)$ and $\tilde{u}(x)$ and then the final solution $u(x)$.

Thus, using examples of the simplest problems of vibrations of a rectilinear bar, it can be shown how the spectral method based on use of the forms and frequencies of free oscillations at the simple homogeneous boundary conditions extends to the case of mixed boundary conditions.

Clearly, this approach can also be used to solve much more complicated problems. Note that the difficulties stem from the choice of coordinate functions for bodies of complex shape. The method of immersing a complex shape body into a simple shape body by adding "fictitious" domains in which supposedly there are no external forces, forces of inertia or friction, and the elasticity modulus is considered to be very small [2] can be applied in such an instance.

6.3 Vibrations of a Three-Dimensional Body (Mixed Boundary Conditions)

Let us imagine a bounded three-dimensional elastic body having a domain (volume) Ω and a smooth surface Γ entirely disposed within the body. Let us draw the orthogonal coordinate system s,t (which defines the surface position of any point on the surface Γ) and place the top of the system of three orthogonal unit vectors $\underline{n},\underline{s},\underline{t}$ at this point. We direct the vectors $\underline{s},\underline{t}$ along the tangents and vector $\underline{n}$ along the normal to the specified surface. Let us assume that the direction of the vectors $\underline{s},\underline{t}$ coincides with the direction of increasing the coordinates s,t, respectively.

Besides the coordinates s, t, let us further introduce the coordinate n, which will be measured along the normal $\underline{n}$ from the point on the surface Γ, such that the coordinates $\underline{n}, \underline{s}, \underline{t}$ form a right-hand coordinate system of coordinates.

In the coordinate system n, s, t we can write the first Eq. (2.13) as

$$-\left(\frac{\partial \underline{N}}{\partial n} + \frac{\partial \underline{S}}{\partial s} + \frac{\partial \underline{T}}{\partial t}\right) + \rho \underline{\ddot{u}} - \underline{p} = 0, \tag{6.56}$$

where $\underline{N}, \underline{S}, \underline{T}$ are vectors of stress acting on the area elements with the normals $\underline{n}, \underline{s}, \underline{t}$. When writing Eq. (6.56) such a formula [1] has been used

$$\mathrm{div}\underline{\underline{\tau}} = \nabla \cdot \underline{\underline{\tau}} = \frac{\partial \underline{N}}{\partial n} + \frac{\partial \underline{S}}{\partial s} + \frac{\partial \underline{T}}{\partial t}, \quad \nabla = \left(\frac{\partial}{\partial n}, \frac{\partial}{\partial s}, \frac{\partial}{\partial t}\right). \tag{6.57}$$

Let us suppose that specific mass $\rho_\Gamma(s, t)$ is distributed on that part of the surface $\Gamma_1 \leq \Gamma$ and that distributed load $\underline{p}_\Gamma(s, t)$ operates. In other words, mass can be concentrated into an infinitely thin layer containing the surface Γ_1

$$\rho(n, s, t) = \rho_\Gamma(s, t)\delta(n - 0)$$

and the load is applied

$$\underline{p}(n, s, t) = \underline{p}_\Gamma(s, t)\delta(n - 0).$$

Formally substituting these expressions for $\rho = \rho(n, s, t)$ and $\underline{p} = \underline{p}(n, s, t)$ into Eq. (6.56), we can integrate the result obtained over n from $-\varepsilon$ to $+\varepsilon$, where ε is a small positive number, and can tend ε to zero

$$\lim_{\varepsilon \to 0} \int_{-\varepsilon}^{+\varepsilon} \left(-\frac{\partial \underline{N}}{\partial n} - \frac{\partial \underline{S}}{\partial s} - \frac{\partial \underline{T}}{\partial t} + \rho \underline{\ddot{u}} - \underline{p}\right) \mathrm{d}n = 0, \tag{6.58}$$

where, in particular

$$\lim_{\varepsilon \to 0} \int_{-\varepsilon}^{+\varepsilon} \frac{\partial \underline{N}}{\partial n} \mathrm{d}n = \lim_{\varepsilon \to 0} (\underline{N}(+\varepsilon, s, t) - \underline{N}(-\varepsilon, s, t)) = \underline{N}(+0, s, t) - \underline{N}(-0, s, t). \tag{6.59}$$

Let us suppose that $\rho(0, s, t)$ and $\underline{p}(0, s, t)$ vary sufficiently smoothly over the surface Γ, such that the derivatives $\frac{\partial \underline{S}}{\partial s}$ and $\frac{\partial \underline{T}}{\partial t}$ would be bounded. Then there will be

$$\lim_{\varepsilon \to 0} \int_{-\varepsilon}^{+\varepsilon} \left(\frac{\partial \underline{S}}{\partial s} + \frac{\partial \underline{T}}{\partial t} \right) dn = 0. \tag{6.60}$$

In addition, according to (6.2)

$$\begin{aligned} &\lim_{\varepsilon \to 0} \int_{-\varepsilon}^{+\varepsilon} \left(\rho_\Gamma(s,t)\delta(n-0)\underline{\ddot{u}}(n,s,t) - \underline{p}_\Gamma(s,t)\delta(n-0) \right) dn \\ &= \rho_\Gamma(s,t)\underline{\ddot{u}}(0,s,t) - \underline{p}_\Gamma(s,t). \end{aligned} \tag{6.61}$$

Considering this formula and formulas (6.59), (6.60), we obtain for the jump of the vector of stresses on the surface Γ_1

$$-\underline{N}(+0,s,t) + \underline{N}(-0,s,t) + \rho_\Gamma(s,t)\underline{\ddot{u}}(0,s,t) - \underline{p}_\Gamma(s,t) = 0. \tag{6.62}$$

Let us further suppose that there is an infinitely thin elastic layer comprising the surface $\Gamma_2 \leq \Gamma(\Gamma_1 + \Gamma_2 = \Gamma)$. The lumped compliance (pliability) of the elastic layer can be represented in the form

$$\underline{\underline{b_\Gamma}}(s,t)\delta(n-0), \tag{6.63}$$

where $\underline{\underline{b_\Gamma}}(s,t)$ is the elasticity tensor at the point of this layer with coordinates s, t.

The lumped compliance (6.63) corresponds to the lumped strain

$$\underline{\underline{\tau}} \cdot \underline{\underline{b_\Gamma}}(s,t)\delta(n-0). \tag{6.64}$$

The expression for the lumped strain of the mentioned layer from external causes can be written as

$$\underline{\underline{e}}_\Gamma(s,t)\delta(n-0), \tag{6.65}$$

where $\underline{\underline{e}}_\Gamma(s,t)$ is the strain tensor from external causes at the point s, t of the surface Γ_2.

Let us rewrite the second Eq. (2.13) in the coordinate system n, s, t, add the terms (6.64) and (6.65), multiply the result obtained by the vector $\underline{n}$ from the left-hand side, integrate the result obtained over the radial coordinate n from $-\varepsilon$ to $+\varepsilon$ and tend ε to zero. This allows us to derive in the limit

$$\int_{-0}^{+0} \frac{1}{2}\underline{n}\cdot(\nabla\underline{u} + (\nabla\underline{u})^*)dn - \int_{-0}^{+0} \left(\underline{N_\Gamma}(s,t) \cdot \underline{\underline{b_\Gamma}}(s,t) + \underline{E_\Gamma}(s,t) \right) \delta(n-0)dn = 0 \tag{6.66}$$

where $\underline{N}_\Gamma(s,t) = \underline{n} \cdot \underline{\underline{\tau}}$ are stresses acting on the area element with the normal $\underline{n}$ distributed over the surface Γ_2; and $\underline{E}_\Gamma(s,t) = \underline{n} \cdot \underline{\underline{e}}_\Gamma(s,t)$ is the strain on the elastic layer containing the surface Γ_2.

We can now represent $\nabla \underline{u}$ as

$$\nabla \underline{u} = \frac{1}{2}(\nabla \underline{u} + (\nabla \underline{u})^*) + \frac{1}{2}(\nabla \underline{u} - (\nabla \underline{u})^*),$$

where the first term represents the tensor of elastic strain, and the second is responsible for the rigid rotation of an infinitely small neighborhood of an arbitrary point of the body [1]. We consider there is no jump in rigid rotation during the transition of the surface Γ_2 along the normal direction. Therefore

$$\int_{-0}^{+0} \frac{1}{2}\underline{n} \cdot (\nabla \underline{u} - (\nabla \underline{u})^*) \mathrm{d}n = 0. \tag{6.67}$$

Summing (6.66) and (6.67), we obtain

$$\int_{-0}^{+0} \Big(\underline{n} \cdot (\nabla \underline{u}) - \Big(\underline{N}_\Gamma(s,t) \cdot \underline{\underline{b_\Gamma}}(s,t) + \underline{E}_\Gamma(s,t)\Big)\delta(n-0)\Big) \mathrm{d}n = 0. \tag{6.68}$$

However

$$\int_{-0}^{+0} (\underline{n} \cdot \nabla \underline{u}) \mathrm{d}n = \underline{u}(+0,s,t) - \underline{u}(-0,s,t). \tag{6.69}$$

So, on the basis of formulas (6.68), (6.69) we arrive at a formula for the jump in displacements on the surface Γ_2

$$\underline{u}(+0,s,t) - \underline{u}(-0,s,t) - \underline{N}_\Gamma(s,t) \cdot \underline{\underline{b_\Gamma}}(s,t) - \underline{E}_\Gamma(s,t) = 0. \tag{6.70}$$

In doing this we considered stress jump (6.62) at continuous displacements on the surface part Γ_1 and displacement jump (6.70) at continuous stresses on the surface part Γ_2, $\Gamma_1 + \Gamma_2 = \Gamma$.

Let us now assume that the surface Γ coincides with the surface of the body. Then the condition (6.62) of the jump in stress $\underline{N}$ turns into the boundary condition on the surface part Γ_1. We set $\underline{N}(+0,s,t) = 0$ in this condition because the elastic body is surrounded by a void, and take the fact that the point $(-0,s,t)$ belongs to the elastic body into account (i.e. $\underline{N}(-0,s,t) = \underline{N}(s,t)$). Then the mixed boundary condition on the surface part Γ_1 in the coordinate system s,t takes the form

$$\underline{N}(s,t) + \rho_\Gamma(s,t)\ddot{\underline{u}}(s,t) - \underline{p}_\Gamma(s,t) = 0. \tag{6.71}$$

On the other hand, there needs to be $\underline{u}(+0,s,t) = 0$ in the condition of the jump in displacement (6.70) on the surface part Γ_2 of an elastic body, since the elastic layer here is based on an absolutely rigid foundation. We should further assume that $\underline{u}(-0,s,t) = \underline{u}(s,t)$ because displacement $\underline{u}(-0,s,t)$ belongs to the elastic body itself. As a result the boundary condition on the surface part Γ_2 in the coordinates s,t is the following

$$-\underline{u}(s,t) - \underline{N}_\Gamma(s,t)\cdot \underline{\underline{b_\Gamma}}(s,t) - \underline{E}_\Gamma(s,t) = 0. \tag{6.72}$$

The coordinate system and designations for forces on the boundary of a body selected in this chapter differ from those in Chap. 5. Furthermore, in Chap. 2 we adopted the three-dimensional Cartesian coordinate system, whereas in this chapter we used the two-dimensional coordinate system mapped on the body surface. However, in the generalized record it is possible to give an identical form to the boundary conditions with the difference that previously simple homogeneous conditions had been assumed, whereas here complex (mixed) boundary nonhomogeneous conditions were considered.

Let us introduce the notation

$$\begin{aligned} X &= \underline{N}(s,t), Y = \underline{u}(s,t),\\ R_\Gamma &= \rho_\Gamma(s,t), p_\Gamma = \underline{p}_\Gamma(s,t),\\ B_\Gamma &= \underline{\underline{b_\Gamma}}(s,t), E_\Gamma = \underline{E}_\Gamma(s,t). \end{aligned} \tag{6.73}$$

Then the boundary conditions (6.71) and (6.72) can be rewritten as the following

$$X + R_\Gamma\ddot{Y} - p_\Gamma = 0 \text{ on } \Gamma_1, \tag{6.74}$$

$$-Y - X\cdot B_\Gamma - E_\Gamma = 0 \text{ on } \Gamma_2 \tag{6.75}$$

Naturally, boundary conditions (6.74) and (6.75) at $R_\Gamma(s,t) = 0,\quad p_\Gamma(s,t) = 0, B_\Gamma(s,t) = 0, E_\Gamma(s,t) = 0$ can be transformed into simple homogeneous conditions

$$X = 0 \text{ on } \Gamma_1 \text{ and } Y = 0 \text{ on } \Gamma_2. \tag{6.76}$$

It is important to mention that the boundary conditions used in this chapter were derived from the equations of dynamic equilibrium and the law of elasticity by means of the notion of the delta function. The mixed variational principle and variational conditions under boundary conditions of the type (6.74) and (6.75) for harmonic oscillations were considered in Chap. 4.

References

1. Lurie, A. I. (2005). *Theory of elasticity* (p. 1050). Berlin: Springer.
2. Sviyazheninov E. D., & Fridman, V. M. (1990). Spectral method for solving the problem of vibrations of a complex geometric-shaped elastic body using fictitious domains. *Proceedings of the U.S.S.R. Academy of Sciences, Mechanics of Solids, 5*, 74–80 (in Russian).

Chapter 7
Exact Solutions of Equations of Oscillation Theory

7.1 Transverse Vibrations of a Rectilinear Beam

Exact solutions of the equations of vibrations of elastic bodies can be obtained, as a rule, for homogeneous bodies with a simple geometry. In constructing exact solutions it is convenient to pass on to equations in terms of displacements, excluding force factors. This chapter provides solutions to some typical problems.

Let us start by considering the free transverse vibrations of a thin bar of constant cross-section (excluding inertia in cross-section rotation and shear strain).

Proceeding from Eq. (1.51), let us write the equation of the free transverse vibrations of a bar in displacements

$$(EIu''(x,t))'' + \rho\ddot{u}(x,t) = 0, \tag{7.1}$$

where x is the coordinate measured along the axis of the bar, $0 \le x \le l$; and l is the length of the bar.

The solution to Eq. (7.1) can be represented in the form

$$u(x,t) = u(x)\sin(\lambda t + \varphi). \tag{7.2}$$

Then, to determine the forms of free oscillations $u = u(x)$ and frequencies of free oscillations λ such equations as

$$(EIu''(x))'' - \lambda^2\rho\, u(x) = 0 \tag{7.3}$$

are valid.

In the case of a bar with a constant cross-section, Eq. (7.3) takes the form

$$u''''(x) - \lambda^2\frac{\rho}{EI}u(x) = 0$$

V. Fridman, *Theory of Elastic Oscillations*, Foundations of Engineering Mechanics, DOI 10.1007/978-981-10-4786-2_7

or

$$u''''(x) - \nu^4 u(x) = 0, \tag{7.4}$$

where

$$\nu^4 = \lambda^2 \frac{\rho}{EI} \tag{7.5}$$

can be denoted.

The solution to Eq. (7.4)

$$u(x) = A_1 \cos \nu x + A_2 \sin \nu x + A_3 \,\mathrm{ch}\, \nu x + A_4 \,\mathrm{sh}\, \nu x$$

can also be written in the form

$$u(x) = C_1 S(\nu x) + C_2 T(\nu x) + C_3 U(\nu x) + C_4 V(\nu x) \tag{7.6}$$

using the following functions of Krylov [3]

$$\begin{aligned} S(\zeta) &= \frac{1}{2}(\mathrm{ch}\,\zeta + \cos\zeta), \quad T(\zeta) = \frac{1}{2}(\mathrm{sh}\,\zeta + \sin\zeta), \\ U(\zeta) &= \frac{1}{2}(\mathrm{ch}\,\zeta - \cos\zeta), \quad V(\zeta) = \frac{1}{2}(\mathrm{sh}\,\zeta - \sin\zeta). \end{aligned} \tag{7.7}$$

The properties of the functions used can be seen in Tables 7.1 and 7.2, where the primes denote derivatives with respect to ζ.

The sequential derivatives of the function $u(x)$ over the variable x have the form

Table 7.1 Successive derivatives of Krylov functions

$f(\zeta)$	$f'(\zeta)$	$f''(\zeta)$	$f'''(\zeta)$	$f''''(\zeta)$
S	V	U	T	S
T	S	V	U	T
U	T	S	V	U
V	U	T	S	V

Table 7.2 The values of Krylov functions and their derivatives at the initial point

$f(\zeta)$	$f(0)$	$f'(0)$	$f''(0)$	$f'''(0)$
S	1	0	0	0
T	0	1	0	0
U	0	0	1	0
V	0	0	0	1

$$
\begin{aligned}
u'(x) &= \nu(C_1V(\nu x) + C_2S(\nu x) + C_3T(\nu x) + C_4U(\nu x)),\\
u''(x) &= \nu^2(C_1U(\nu x) + C_2V(\nu x) + C_3S(\nu x) + C_4T(\nu x)),\\
u'''(x) &= \nu^3(C_1T(\nu x) + C_2U(\nu x) + C_3V(\nu x) + C_4S(\nu x)).
\end{aligned}
\tag{7.8}
$$

The constants $C_i, i = 1, 2, 3, 4$, in solution (7.6) can be determined from the boundary conditions of the problem.

To compile the equation of frequencies, it is necessary using (7.6) and (7.8) to express the boundary conditions of the problem through the functions of Krylov, and then to record the condition of existence of nonzero solutions for the constants C_i.

Example 1 Let us consider the problem of determining the frequencies and forms of free vibrations of a beam simply (freely) supported at the ends. In this case the boundary conditions are

$$
\begin{aligned}
u(0) = 0, \quad EIu''(0) = 0,\\
u(l) = 0, \quad EIu''(l) = 0.
\end{aligned}
\tag{7.9}
$$

From the first condition (7.9) it follows that $C_1 = 0$ in the solution (7.6). From the second condition (7.9) at $x = 0$ we obtain, using the expression for $u''(0)$ from (7.8)

$$\nu^2C_3 = 0. \tag{7.10}$$

Since the trivial solution $u(x) \equiv 0$ corresponds to the value $\nu = 0$ ($\lambda = 0$) at the boundary conditions given, $C_3 = 0$ needs to be in (7.10).

From the boundary conditions (7.9) at $x = l$ and the expressions (7.6), as well as (7.8) for $u''(l)$, we obtain two homogeneous equations with respect to the coefficients C_2, C_4

$$
\begin{aligned}
C_2T(\alpha) + C_4V(\alpha) = 0,\\
C_2V(\alpha) + C_4T(\alpha) = 0,
\end{aligned}
\tag{7.11}
$$

where the dimensionless parameter

$$\alpha = \nu l \tag{7.12}$$

is introduced.

From the condition of existence of a nonzero solution of the system (7.11), we obtain an equation to determine the frequencies of free oscillations

$$T^2(\alpha) - V^2(\alpha) = 0$$

or after transformations

$$\sin(\alpha) = 0, \tag{7.13}$$

whose roots are

$$\alpha_i = i\pi, \quad i = 1, 2, \ldots \tag{7.14}$$

As a result, using (7.5), (7.6), (7.11), (7.12) and (7.14), we obtain the following expressions for the forms and frequencies of free oscillations (natural forms and frequencies) of a freely (simply) supported beam

$$u_i(x) = A_i \sin\frac{i\pi x}{l}, \quad \lambda_i = \left(\frac{i\pi}{l}\right)^2 \sqrt{\frac{EI}{\rho}}, \quad i = 1, 2, \ldots \tag{7.15}$$

By normalizing the forms of free oscillations we can facilitate solutions to the problems of the forced oscillations of the beams; that is, the coefficients A_i can be determined from the condition

$$\int_0^l \rho u_i^2(x) \mathrm{d}x = 1, \quad \text{from which} \quad A_i = \sqrt{\frac{2}{\rho l}}. \tag{7.16}$$

Example 2 Let us consider the problem of determining the frequencies and forms of free oscillations of a beam with sliding sealing at the ends.

The boundary conditions have the form

$$\begin{aligned} u'(0) &= 0, \quad EIu'''(0) = 0, \\ u'(l) &= 0, \quad EIu'''(l) = 0. \end{aligned} \tag{7.17}$$

Substituting expressions (7.8) into (7.17), we obtain the system of equations

$$\begin{gathered} \nu C_2 = 0, \quad \nu^3 C_4 = 0, \\ \nu(C_1 V(\alpha) + C_2 S(\alpha) + C_3 T(\alpha) + C_4 U(\alpha)) = 0, \\ \nu^3(C_1 T(\alpha) + C_2 U(\alpha) + C_3 V(\alpha) + C_4 S(\alpha)) = 0, \end{gathered} \tag{7.18}$$

having the zero root

$$\nu_1 = 0 \quad \text{or} \quad \lambda_1 = 0, \tag{7.19}$$

to which there corresponds the movement of the beam as a rigid whole

$$u_0(x) = A_0, \quad A_0 = \sqrt{\frac{1}{\rho l}}. \tag{7.20}$$

Recall that $\alpha = \nu l$ is the dimensionless parameter.

To the nonzero frequencies there corresponds the solution

$$\begin{gathered} C_2 = 0, \quad C_4 = 0, \\ C_1 V(\alpha) + C_3 T(\alpha) = 0, \\ C_1 T(\alpha) + C_3 V(\alpha) = 0. \end{gathered} \tag{7.21}$$

The condition of existence of the nonzero solution for C_1 and C_3 is

$$V^2(\alpha) - T^2(\alpha) = 0$$

or

$$\sin(\alpha) = 0, \tag{7.22}$$

whose roots

$$\alpha_i = i\pi, \quad i = 1, 2, \ldots \tag{7.23}$$

have the same form as in the example of free oscillations of the beam with simple support conditions.

Finally, for the normalized forms and frequencies of free vibrations of the beam with the boundary conditions (7.17), based on (7.5), (7.6), (7.20), (7.21) and (7.23), we obtain

$$\begin{gathered} u_0(x) = \sqrt{\tfrac{1}{\rho l}}, \quad \lambda_0 = 0, \\ u_i(x) = \sqrt{\tfrac{2}{\rho l}} \cos\tfrac{i\pi x}{l}, \quad \lambda_i = \left(\tfrac{i\pi}{l}\right)^2 \sqrt{\tfrac{EI}{\rho}}, \quad i = 1, 2, \ldots \end{gathered} \tag{7.24}$$

Example 3 Let us consider the problem of determining the frequencies and forms of free oscillations of a beam with fixed ends.

The boundary conditions have the form

$$u(0) = 0, \quad u'(0) = 0, \quad u(l) = 0, \quad u'(l) = 0. \tag{7.25}$$

From conditions (7.25) and expressions (7.6) and (7.8) we have

$$\begin{gathered} C_1 = 0, \quad C_2 = 0, \\ C_3 U(\alpha) + C_4 V(\alpha) = 0, \\ C_3 T(\alpha) + C_4 U(\alpha) = 0. \end{gathered} \tag{7.26}$$

The condition of existence of the nonzero solutions for C_3 and C_4 is

$$U^2(\alpha) - T(\alpha) V(\alpha) = 0. \tag{7.27}$$

Substituting expressions (7.7) for the functions of Krylov, we obtain the transcendental equation with respect to α

$$\operatorname{ch}\alpha\cos\alpha = 1. \tag{7.28}$$

The roots α_i of Eq. (7.28) can be computed approximately equal to

$$\alpha_1 = 4730, \quad \alpha_2 = 7859, \quad \alpha_3 = 10{,}996, \ldots \tag{7.29}$$

and the corresponding frequencies of free oscillations, according to (7.5) and (7.12), can be determined by the formula

$$\lambda_i = \left(\frac{\alpha_i}{l}\right)^2 \sqrt{\frac{EI}{\rho}}, \quad i = 1, 2, \ldots \tag{7.30}$$

The forms of free oscillations in this case are determined to be more complicated than in Examples 1, 2 and, in view of (7.26), (7.6), take the following form

$$u_i(x) = A_i(V(\alpha_i)U(\nu_i x) - U(\alpha_i)V(\nu_i x)) \tag{7.31}$$

or

$$\begin{aligned} u_i(x) = A_i((&\operatorname{sh}\alpha_i - \sin\alpha_i)(\operatorname{ch}\nu_i x - \cos\nu_i x) \\ &-(\operatorname{ch}\alpha_i - \cos\alpha_i)(\operatorname{sh}\nu_i x - \sin\nu_i x)). \end{aligned} \tag{7.32}$$

Using direct substitution to determine A_i from the normalization condition (7.16) of the forms of free oscillations, solution (7.32) is rather laborious.

Calculation simplification can be eased using formula (7.33), which allows determining the square of the norm of the form of free oscillations by the values of this function and its derivatives at the ends of the beam, namely

$$\int_0^l \rho u_i^2(x)\mathrm{d}x = \frac{\rho l}{4}\left(\left(\frac{\mathrm{d}^2 u_i}{\mathrm{d}\zeta^2}\right)^2 + u_i^2 - 2\frac{\mathrm{d}u_i}{\mathrm{d}\zeta}\frac{\mathrm{d}^3 u_i}{\mathrm{d}\zeta^3}\right)_{\zeta=\alpha_i} = 1, \tag{7.33}$$

where the dimensionless coordinate has been introduced

$$\zeta = \frac{\alpha}{l}x = \nu x. \tag{7.34}$$

The condition $\zeta = \alpha_i$ can be replaced by $x = l, \lambda = \lambda_i$.

By virtue of the boundary conditions (7.25) at $x = l$ we have

$$u_i|_{\zeta=\alpha_i} = 0, \quad \frac{\mathrm{d}u_i}{\mathrm{d}\zeta}|_{\zeta=\alpha_i} = 0.$$

Therefore

$$\int_0^l \rho u_i^2(x)\mathrm{d}x = \frac{\rho\, l}{4}\left(\frac{\mathrm{d}^2 u_i}{\mathrm{d}\zeta^2}\right)^2_{\zeta=\alpha_i} = \frac{\rho\, l}{4} A_i^2 (V(\alpha_i)S(\alpha_i) - U(\alpha_i)T(\alpha_i))^2 = 1,$$

where

$$A_i = \frac{2}{\sqrt{\rho l}} \frac{1}{|V(\alpha_i)S(\alpha_i) - U(\alpha_i)T(\alpha_i)|} \tag{7.35}$$

or, considering expressions (7.7)

$$A_i = \frac{4}{\sqrt{\rho l}} \frac{1}{|\mathrm{sh}\, \alpha_i \, \cos \alpha_i - \mathrm{ch}\, \alpha_i \sin \alpha_i|}. \tag{7.36}$$

Substituting (7.36) into (7.32), we obtain an expression for the normalized ith form of free oscillations of the beam with fixed edges.

As is evident from (7.32), solution of the problem in the case of the boundary conditions (7.25) can be expressed through the trigonometric and hyperbolic functions of the argument.

Let us now consider the method of normalization of forms of free bending vibrations of a beam by their values on the boundaries. Returning to the operator record of the equations of free bending vibrations of a beam (see Chap. 3), we note there were

$$\begin{aligned} &\int_0^l N\eta_i \eta_j \mathrm{d}x - \lambda_i^2 \int_0^l R\eta_i \eta_j \mathrm{d}x = 0, \\ &\int_0^l N\eta_j \eta_i \mathrm{d}x - \lambda_j^2 \int_0^l R\eta_j \eta_i \mathrm{d}x = 0. \end{aligned} \tag{7.37}$$

Representing the operator N as the product of three operators and using the conjugacy properties of the operators D and D^* in the sense of Lagrange, we obtain

$$\begin{aligned} \int_0^l N\eta_i\eta_j \mathrm{d}x &= \int_0^l DB^{-1}D^*\eta_i\eta_j \mathrm{d}x = \int_0^l D\xi_i\eta_j \mathrm{d}x \\ &= \int_0^l \xi_i D^*\eta_j \mathrm{d}x - \xi_i\eta_j|_{x=0}^{x=l} = \int_0^l \xi_i B\xi_j \mathrm{d}x - \xi_i\eta_j|_{x=0}^{x=l}. \end{aligned}$$

Analogously

$$\int_0^l N\eta_j\eta_i dx = \int_0^l \xi_j B\xi_i dx - \xi_j\eta_i|_{x=0}^{x=l}.$$

Subtracting the first equation from the second equation of (7.37) and considering the properties of self-adjointness of the operators R and B, we obtain

$$(\lambda_i^2 - \lambda_j^2)\int_0^l R\eta_i\eta_j dx + (\xi_i\eta_j - \xi_j\eta_i)|_{x=0}^{x=l} = 0,$$

where

$$\int_0^l R\,\eta_i\eta_j dx = \frac{(\xi_j\eta_i - \xi_i\eta_j)|_{x=0}^{x=l}}{(\lambda_i^2 - \lambda_j^2)}. \tag{7.38}$$

In the event of simple homogeneous boundary conditions we have

$$\xi_i\eta_j|_{x=0}^{x=l} = (Q_i\,u_j + M_i\,\vartheta_j)|_{x=0}^{x=l} = 0,$$
$$\xi_j\eta_i|_{x=0}^{x=l} = (Q_j\,u_i + M_j\,\vartheta_i)|_{x=0}^{x=l} = 0.$$

Then, at $\lambda_i^2 \neq \lambda_j^2$

$$\int_0^l R\eta_i\eta_j dx = 0, \tag{7.39}$$

(i.e. the forms of free oscillations are orthogonal), and at $\lambda_i^2 = \lambda_j^2$

$$\int_0^l R\eta_i\eta_i dx = \frac{0}{0}$$

we get uncertainty.

To disclose this uncertainty (i.e. to determine the norm of the ith form of free oscillations), we proceed as follows.

Suppose that $\lambda^2 \neq \lambda_i^2$ (i.e. $\lambda^2 = \lambda_i^2 + \delta$, where δ is a small number) and η, ξ are the solution to the problem

$$N\eta - \lambda^2 R\eta = 0,$$

satisfying the same homogeneous conditions at the left end $\xi\,\eta|_{x=0} = 0$ of the rod as the functions ξ_i, η_i. However, in this case $\xi\,\eta|_{x=l} \neq 0$. Then

$$\int_0^l R\eta_i\,\eta\mathrm{d}x = \frac{(\xi\,\eta_i - \xi_i\eta)|_{x=l}}{(\lambda_i^2 - \lambda^2)}$$

The squared norm of the *i*th form of free oscillations can be found from the limiting relations using the rule of L'Hôpital

$$\int_0^l R\eta_i\eta_i\mathrm{d}x = \lim_{\lambda^2 \to \lambda_i^2} \frac{(\xi\,\eta_i - \xi_i\eta)|_{x=l}}{(\lambda_i^2 - \lambda^2)} = \frac{\frac{\mathrm{d}}{\mathrm{d}\lambda}(\xi\,\eta_i - \xi_i\eta)|_{x=l,\lambda=\lambda_i}}{-2\,\lambda_i} \tag{7.40}$$

or

$$\int_0^l R\eta_i\eta_i\mathrm{d}x = \frac{\left(\frac{\mathrm{d}Q}{\mathrm{d}\lambda}u_i + \frac{\mathrm{d}M}{\mathrm{d}\lambda}\vartheta_i - Q_i\frac{\mathrm{d}u}{\mathrm{d}\lambda} - M_i\frac{\mathrm{d}\vartheta}{\mathrm{d}\lambda}\right)_{x=l,\lambda=\lambda_i}}{-2\lambda_i}. \tag{7.41}$$

Since the bending vibrations of a thin beam with a constant cross-section by virtue of (1.44) are

$$\vartheta = -u'; \quad M = -EJu''; \quad Q = -EJu''',$$

where the prime means differentiation with respect to x, instead of (7.41) we obtain

$$\int_0^l R\,\eta_i\eta_i\mathrm{d}x$$
$$= \frac{EJ\left(-\frac{\mathrm{d}u'''}{\mathrm{d}\lambda}u_i + \frac{\mathrm{d}u''}{d\lambda}u_i' + u_i'''\frac{\mathrm{d}u}{\mathrm{d}\lambda} - u_i''\frac{\mathrm{d}u'}{d\lambda}\right)_{x=l,\lambda=\lambda_i}}{-2\lambda_i}. \tag{7.42}$$

Should we wish to express the derivatives with respect to λ and x through the derivatives on the dimensionless variable ζ, then taking relations (7.5), (7.12), (7.34) as well as Eq. (7.4) into account, we have

$$\frac{\mathrm{d}}{\mathrm{d}\lambda}=\frac{\mathrm{d}}{\mathrm{d}\zeta}\frac{\mathrm{d}\zeta}{\mathrm{d}\alpha}\frac{\mathrm{d}\alpha}{\mathrm{d}\lambda}=\frac{x}{2l}\frac{\alpha}{\lambda}\frac{\mathrm{d}}{\mathrm{d}\zeta},$$
$$\frac{\mathrm{d}}{\mathrm{d}x}=\frac{\mathrm{d}}{\mathrm{d}\zeta}\frac{\mathrm{d}\zeta}{\mathrm{d}x}=\frac{\alpha}{l}\frac{\mathrm{d}}{\mathrm{d}\zeta},\quad \frac{\mathrm{d}^2}{\mathrm{d}x^2}=\left(\frac{\alpha}{l}\right)^2\frac{\mathrm{d}^2}{\mathrm{d}\zeta^2},$$
$$\frac{\mathrm{d}^3}{\mathrm{d}x^3}=\left(\frac{\alpha}{l}\right)^3\frac{\mathrm{d}^3}{\mathrm{d}\zeta^3},\quad \frac{\mathrm{d}^4u_i}{\mathrm{d}\zeta^4}=u_i.$$

Then, instead of (7.42), we finally obtain a formula that allows the forms of free bending vibrations of a thin beam with a constant cross-section to be normalized, and the values of these functions and their derivatives at the border to be known

$$\int_0^l R\eta_i\eta_i\mathrm{d}x=\frac{\rho l}{4}\left(u_i^2-2\frac{\mathrm{d}u_i}{\mathrm{d}\zeta}\frac{\mathrm{d}^3u_i}{\mathrm{d}\zeta^3}+\left(\frac{\mathrm{d}^2u_i}{\mathrm{d}\zeta^2}\right)^2\right)_{x=l,\lambda=\lambda_i}=1.$$

Let us now consider the forced harmonic oscillations of a thin beam with a constant cross-section and dwell on the case when both ends of the beam are fixed.

Let the external distributed harmonic force act on the beam with

$$f(x,t)=f(x)\cos(\lambda t+\alpha), \tag{7.43}$$

whose amplitude is

$$f(x)=\begin{pmatrix}p(x)\\0\end{pmatrix}.$$

The problem can be reduced to finding the solution of the equation, which we write in the amplitudes of displacements $\eta(x)$

$$N\eta-\lambda^2R\eta-f=0 \tag{7.44}$$

at the boundary conditions

$$\eta(0)=0,\quad \eta(l)=0, \tag{7.45}$$

where $\eta=\eta(x)=\begin{pmatrix}u(x)\\ \vartheta(x)\end{pmatrix}$ (designations as given in Chap. 1).

The solution of the problem (7.44) and (7.45) can be constructed in various ways.

1. The first way consists in expanding the solution to a series of the forms of free oscillations of a beam with the same boundary conditions (7.45) as in the original case.

In this particular problem the forms of free oscillations of the beam can be viewed as in (7.32), expressed in terms of the trigonometric and hyperbolic functions of the argument and the natural frequencies λ_i determined from the transcendental equation (7.28).

In this case the solution of the problem (7.44) and (7.45) can be written as

$$\eta(x) = \sum_{i=1}^{n} \beta_i \eta_i(x), \tag{7.46}$$

where the forms of free oscillations

$$\eta_i(x) = \begin{pmatrix} u_i(x) \\ \vartheta_i(x) \end{pmatrix}, \quad \vartheta_i(x) = -u_i'(x)$$

and frequencies λ_i can be found from the equations and boundary conditions

$$N\eta_i - \lambda_i^2 R\eta_i = 0, \quad \eta_i(0) = 0, \quad \eta_i(l) = 0. \tag{7.47}$$

Hence, the coefficients β_i are equal to (see Chap. 3)

$$\beta_i = \frac{f_i}{\lambda_i^2 - \lambda^2}, \quad f_i = \int_0^l f(x)\eta_i(x)\mathrm{d}x, \quad i = 1, 2, \ldots, n. \tag{7.48}$$

Having obtained the amplitude of displacement $\eta(x)$ from (7.46), we can find the amplitude of corresponding internal forces $\xi(x)$ on the basis of the relations

$$\xi = B^{-1}D^*\eta. \tag{7.49}$$

2. The second way to solve the problem of forced beam oscillations with the conditions (7.45) is to use the forms of free oscillations of a bar with simplified boundary conditions. As the latter there can be used the forms (7.15) of free oscillations of a beam with the simple support conditions of its ends, which can be expressed in terms of trigonometric functions of the argument, and for the frequencies of free oscillations there obtained analytical expressions.

Let us write the equations for the amplitudes of forces $\xi(x)$ and displacements $\eta(x)$

$$D\xi - \lambda^2 R\eta - f = 0, \quad D^*\eta - B\xi = 0 \tag{7.50}$$

with the boundary conditions (7.45). We seek a solution in the form

$$\eta(x) = \tilde{\eta}(x), \quad \xi(x) = \tilde{\xi}(x) + \widehat{\xi}(x), \tag{7.51}$$

where

$$\begin{aligned} \tilde{\eta}(x) &= \sum_{i=1}^{n} \beta_i \tilde{\eta}_i(x), \\ \tilde{\xi}(x) &= \sum_{i=1}^{n} \alpha_i \tilde{\xi}_i(x), \quad \widehat{\xi}(x) = \sum_{j=1}^{2} a_j \widehat{\xi}_j(x), \end{aligned} \tag{7.52}$$

and $\tilde{\eta}_i(x)$, $\tilde{\xi}_i(x)$ are the solutions of the equations

$$D\tilde{\xi}_i - \tilde{\lambda}_i^2 R\tilde{\eta}_i = 0, \quad D^* \tilde{\eta}_i - B\tilde{\xi}_i = 0, \tag{7.53}$$

at the conditions

$$\tilde{u}_i(0) = \tilde{u}_i(l) = 0, \quad \tilde{M}_i(0) = \tilde{M}_i(l) = 0 \tag{7.54}$$

having the form

$$\begin{aligned} \tilde{\eta}_i(x) &= \begin{pmatrix} \tilde{u}_i \\ \tilde{\vartheta}_i \end{pmatrix} = A_i \begin{pmatrix} \sin \frac{i\pi x}{l} \\ -\frac{i\pi}{l} \cos \frac{i\pi x}{l} \end{pmatrix}, \\ \tilde{\xi}_i(x) &= \begin{pmatrix} \tilde{Q}_i \\ \tilde{M}_i \end{pmatrix} = A_i EJ \begin{pmatrix} \left(\frac{i\pi}{l}\right)^3 \cos \frac{i\pi x}{l} \\ \left(\frac{i\pi}{l}\right)^2 \sin \frac{i\pi x}{l} \end{pmatrix}, \\ A_i &= \sqrt{\frac{2}{\rho l}}, \quad \lambda_i = \left(\frac{i\pi}{l}\right)^2 \sqrt{\frac{EJ}{\rho}}, \end{aligned} \tag{7.55}$$

and as $\widehat{\xi}_j(x)$, $j = 1, 2$ we choose the functions

$$\widehat{\xi}_1 = \begin{pmatrix} -\frac{1}{l} \\ 1 - \frac{x}{l} \end{pmatrix}, \quad \widehat{\xi}_2 = \begin{pmatrix} \frac{1}{l} \\ \frac{x}{l} \end{pmatrix}. \tag{7.56}$$

Hence, the conditions on displacements at the ends will be satisfied; that is

$$u(0) = 0, \; u(l) = 0,$$

while the conditions on the angles $\vartheta(0)$ and $\vartheta(l)$ can be fulfilled by selecting the coefficients of the expansions (7.52).

To determine the unknowns α_i, β_i, $i = 1, 2, \ldots, n$, a_j, $j = 1, 2$, we can write the system of projection conditions as

$$\int_0^l (D\xi - \lambda^2 R\eta - f)\tilde{\eta}_i \mathrm{d}x = 0,$$
$$\int_0^l (D^*\eta - B\xi)\tilde{\xi}_i \mathrm{d}x = 0. \tag{7.57}$$

Taking into account the transformations (7.57), the expansions (7.52), the properties of adjointness of the operators D, D^* and self-adjointness of the operators B, R, we arrive at a system of algebraic equations for the coefficients entered

$$\tilde{\lambda}_i^2 \alpha_i - \lambda^2 \beta_i + \sum_{j=1}^{2} a_j (B_{ji} + C_{ji}) - \tilde{f}_i = 0,$$
$$\tilde{\lambda}_i^2 \beta_i - \tilde{\lambda}_i^2 \alpha_i - \sum_{j=1}^{2} a_j B_{ji} = 0, \quad i = 1, 2, \ldots, n. \tag{7.58}$$

where the following notation is introduced

$$\tilde{f}_i = \int_0^l f\tilde{\eta}_i \mathrm{d}x, \quad B_{ji} = \int_0^l \hat{\xi}_j B \tilde{\xi}_i \mathrm{d}x,$$
$$C_{1i} = -\frac{i\pi}{l} A_i, \quad C_{2i} = (-1)^{i+1} C_{1i} \tag{7.59}$$

As a consequence of Eq. (7.58), we can express α_i, β_i as

$$\beta_i = \frac{1}{\tilde{\lambda}_i^2 - \lambda^2}\left(\tilde{f}_i - \sum_{j=1}^{2} a_j C_{ji}\right), \quad \alpha_i = \beta_i - \frac{1}{\tilde{\lambda}_i^2} \sum_{j=1}^{2} a_j B_{ji}. \tag{7.60}$$

There remain two boundary conditions on the angles $\vartheta(x)$ at the ends of the rod that need to be fulfilled to determine a_j.

Let us fulfill the boundary conditions on $\vartheta(x)$ integrally; that is, write the equation as

$$\vartheta' - \beta M = 0,$$

where β, as before, is the distributed compliance at bending. Let us multiply it sequentially by $\hat{M}_1(x) = 1 - \frac{x}{l}, \hat{M}_2(x) = \frac{x}{l}$ and integrate it over the length. We get

$$\int_0^l (\vartheta' - \beta M)\widehat{M}_j \mathrm{d}x = 0, \quad j = 1, 2. \tag{7.61}$$

since

$$\int_0^l \vartheta' \widehat{M}_j \mathrm{d}x = \widehat{M}_j(x)\vartheta(x)|_{x=0}^{x=l} - \int_0^l \vartheta \widehat{M}_j' \mathrm{d}x$$
$$= \widehat{M}_j(l)\vartheta(l) - \widehat{M}_j(0)\vartheta(0) + \widehat{M}_j'(u(l) - u(0)) = 0.$$

We used the boundary conditions (7.45) in this equation and the fact that $\widehat{M}_j' = \mathrm{const}(j)$.

Then, at $j = 1$ and $j = 2$, respectively, instead of (7.61) we get

$$\begin{aligned} \sum_{i=1}^{n} \alpha_i B_{1i} + \sum_{k=1}^{2} a_k \widehat{B}_{k1} = 0, \\ \sum_{i=1}^{n} \alpha_i B_{2i} + \sum_{k=1}^{2} a_k \widehat{B}_{k2} = 0, \end{aligned} \tag{7.62}$$

where

$$\widehat{B}_{kj} = \int_0^l \beta \widehat{M}_k \widehat{M}_j \mathrm{d}x, \quad k, j = 1, 2.$$

Substituting α_i into (7.62), expressed by means of (7.60) through $a_j, j = 1, 2$, we obtain a system of two algebraic equations that can be recorded in matrix form

$$Ay = b, \tag{7.63}$$

where $y = (a_1, a_2)$ is the column vector of the unknown coefficients; $A = A(\lambda^2)$ is a matrix whose elements depend on the known frequency of external action λ; and b is the known vector of the right-hand sides.

The solution to (7.63)

$$y = A^{-1}b$$

determines the unknowns a_j, $j = 1, 2$. Further, with the help of relations (7.60), (7.55), (7.56), (7.52) and (7.51), a solution to the original problem of the forced vibrations of a beam can be constructed.

Using this approach the problem of free oscillations can be solved (i.e. the problem of defining the forms and frequencies of free oscillations of a bar with fixed ends). In this case the solution of Eq. (7.50) should be regarded as $f(x) = 0$, while the frequency of free oscillations λ should be regarded as unknown. Then, instead of (7.63), we obtain the system of homogeneous equations

$$Ay = 0, \tag{7.64}$$

from the condition of existence of the nonzero solution in which $|A(\lambda^2)| = 0$, we obtain an equation to determine the unknown frequencies λ_i of the free oscillations of a bar with clamped ends; whereas the solution to Eq. (7.64), together with expressions (7.60) and expansions (7.51) and (7.52), gives the forms of the free oscillations of a beam.

Note that when the problem of free oscillations of a beam with a constant cross-section is solved taking such factors as shear strain and inertia of rotation of the cross-section into account, the solution for the forms of free oscillations of the beam with freely supported ends has exactly the same form (7.15) as in the simpler case. At the same time, expressing the forms of free oscillations of the beam with clamped edges is difficult since these solutions depend on the coefficients of the equation and the relations between them. The problem of the forced oscillations of a bar with clamped edges can be solved by a much simpler second method based on using the forms of the free oscillations of a beam with simply supported edges.

7.2 Free Oscillations of a Ring of Uniform Cross-Section

Let us consider the problem of the vibrations of a ring of constant cross-section in the ring plane. The corresponding equations have been presented above and have the form (2.36)–(2.38).

In the case of the free oscillations of a ring, these equations include the acting external forces and dislocation strains equal to zero

$$p^1 = 0, \quad p^2 = 0, \quad m = 0, \quad g_1 = 0, \quad g_2 = 0, \quad k = 0. \tag{7.65}$$

Therefore, we are searching for a solution to Eqs. (2.36)–(2.38) in the form

$$\xi(x,t) = \xi(x)\sin\lambda t, \quad \eta(x,t) = \eta(x)\sin\lambda t, \tag{7.66}$$

and for the amplitude values of the components of generalized forces $\xi(x) = \begin{pmatrix} Q^1(x) \\ Q^2(x) \\ M(x) \end{pmatrix}$ and generalized displacements $\eta(x) = \begin{pmatrix} u_1(x) \\ u_2(x) \\ \vartheta(x) \end{pmatrix}$, we obtain the system of equations

$$\begin{aligned} -\frac{dQ^1}{dx} - \frac{Q^2}{R} - \lambda^2 \rho u_1 &= 0, \\ \frac{du_1}{dx} + \frac{u_2}{R} - b_1 Q^1 &= 0, \end{aligned} \tag{7.67}$$

$$\begin{aligned} &-\frac{dQ^2}{dx} + \frac{Q^1}{R} - \lambda^2 \rho u_2 = 0, \\ &Q^2 - \frac{dM}{dx} - \lambda^2 j \vartheta = 0, \\ &\frac{du_2}{dx} - \frac{u_1}{R} + \vartheta - b_2 Q^2 = 0, \\ &\frac{d\vartheta}{dx} - \beta M = 0. \end{aligned} \tag{7.68}$$

Should we want to go from the circumferential coordinate x to the angular coordinate φ in Eqs. (7.67) and (7.68), we need to define the position of a point on the ring

$$x = R\varphi, \quad \frac{d}{dx} = \frac{1}{R}\frac{d}{d\varphi}, \tag{7.69}$$

then, as far as the forces $\xi(\varphi) = \begin{pmatrix} Q^1(\varphi) \\ Q^2(\varphi) \\ M(\varphi) \end{pmatrix}$ and displacements $\eta(\varphi) = \begin{pmatrix} u_1(\varphi) \\ u_2(\varphi) \\ \vartheta(\varphi) \end{pmatrix}$ are concerned, we arrive at the system of equations

$$\begin{aligned} -\frac{1}{R}\frac{dQ^1}{d\varphi} - \frac{Q^2}{R} - \lambda^2 \rho\, u_1 &= 0, \\ \frac{1}{R}\frac{du_1}{d\varphi} + \frac{u_2}{R} - b_1 Q^1 &= 0, \end{aligned} \tag{7.70}$$

$$\begin{aligned} &-\frac{1}{R}\frac{dQ^2}{d\varphi} + \frac{Q^1}{R} - \lambda^2 \rho\, u_2 = 0, \\ &Q^2 - \frac{1}{R}\frac{dM}{d\varphi} - \lambda^2 j\, \vartheta = 0, \\ &\frac{1}{R}\frac{du_2}{d\varphi} - \frac{u_1}{R} + \vartheta - b_2 Q^2 = 0, \\ &\frac{1}{R}\frac{d\vartheta}{d\varphi} - \beta M = 0. \end{aligned} \tag{7.71}$$

The solutions of Eqs. (7.70) and (7.71), satisfying the conditions of periodicity

$$\xi(0) = \xi(2\pi), \quad \eta(0) = \eta(2\pi) \tag{7.72}$$

and symmetrical relative to the horizontal axis, can be written as follows

$$\begin{aligned}
\xi(\varphi) &= \begin{pmatrix} Q^1(\varphi) \\ Q^2(\varphi) \\ M(\varphi) \end{pmatrix} = \begin{pmatrix} Q^1 \cos i\varphi \\ Q^2 \sin i\varphi \\ M \cos i\varphi \end{pmatrix}, \\
\eta(\varphi) &= \begin{pmatrix} u_1(\varphi) \\ u_2(\varphi) \\ \vartheta(\varphi) \end{pmatrix} = \begin{pmatrix} u_1 \sin i\varphi \\ u_2 \cos i\varphi \\ \vartheta \sin i\varphi \end{pmatrix}, \quad i = 1, 2, \ldots
\end{aligned} \tag{7.73}$$

Then, for the forces Q^1, Q^2, M and displacements u_1, u_2, ϑ, we obtain the system of homogeneous algebraic equations

$$\begin{aligned}
&\frac{1}{R} i Q^1 - \frac{Q^2}{R} - \lambda^2 \rho u_1 = 0, \\
&\frac{1}{R} i u_1 + \frac{u_2}{R} - b_1 Q^1 = 0,
\end{aligned} \tag{7.74}$$

$$\begin{aligned}
&-\frac{1}{R} i Q^2 + \frac{Q^1}{R} - \lambda^2 \rho u_2 = 0, \\
&Q^2 + \frac{1}{R} i M - \lambda^2 j \vartheta = 0, \\
&-\frac{1}{R} i u_2 - \frac{u_1}{R} + \vartheta - b_2 Q^2 = 0, \\
&\frac{1}{R} i \vartheta - \beta M = 0.
\end{aligned} \tag{7.75}$$

The system of six algebraic equations (7.74) and (7.75) can be written in matrix form

$$A z = 0, \tag{7.76}$$

where the unknown $z = (Q^1, Q^2, M, u_1, u_2, \vartheta)$ has been introduced, and $A = A(\lambda^2)$ is a 6×6 matrix. From here on we do not differentiate between the recording of row vectors and column vectors.

The frequencies of the free oscillations of the ring $\lambda = \lambda(i)$ can be determined from the existence condition of a nonzero solution of the system (7.76)

$$|A(\lambda^2)| = 0, \tag{7.77}$$

while the forms of free oscillations that correspond to them are determined by subsequent solution of the system of Eqs. (7.76).

In the general case of the free oscillations of a ring, condition (7.77) leads us to an equation of the third degree about λ^2

$$\begin{gathered}\lambda^6(R^6 b_1 b_2 \beta\,\rho^2 j)\\ -\lambda^4\left((i^2+1)(R^4 b_2\beta\,\rho j + R^4 b_1 \beta\,\rho j) + R^4 b_1 b_2 \rho^2 i^2 + R^6 b_1 \beta\,\rho^2\right)\\ +\lambda^2\left((i^2+1)(R^4\beta\,\rho + R^2 b_2 \rho\, i^2 + R^2 b_1 \rho\, i^2) + R^2\beta j(i^2-1)^2\right)\\ -i^2(i^2-1)^2 = 0.\end{gathered}$$

If we set the shear deformation and rotation inertia of the cross-section of a ring equal to zero ($b_2 = 0, j = 0$), then the ring can take a biquadratic equation for λ^2

$$\lambda^4(R^6 b_1 \beta\,\rho^2) - \lambda^2(i^2+1)(R^4\beta\,\rho + R^2 b_1 \rho\, i^2) + i^2(i^2-1)^2 = 0.$$

Let us consider the simplest case, which is the free oscillations of an inextensible ring assuming that its cross-sectional dimensions are small compared with the radius R of the centerline. Making these assumptions in Eqs. (7.70), (7.71), we can neglect shear deformation, rotational inertia of the cross-section of the ring and its pliability on stretching

$$b_2 = 0, \quad j = 0, \quad b_1 = 0. \tag{7.78}$$

In this case condition (7.77) reduces to the equation

$$(1-i^2)^2 - \lambda^2 \rho R^4 \beta\left(1 + \frac{1}{i^2}\right) = 0, \tag{7.79}$$

from which we obtain the natural frequency of the ring for the ith natural form of vibrations [3]

$$\lambda_i = \sqrt{\frac{i^2(1-i^2)^2}{\rho R^4 \beta(1+i^2)}}. \tag{7.80}$$

At $i = 1$ we get $\lambda_1 = 0$. In this case

$$\xi(\varphi) = \begin{pmatrix} Q^1\cos\varphi \\ Q^2\sin\varphi \\ M\cos\varphi \end{pmatrix}, \quad \eta(\varphi) = \begin{pmatrix} u_1\sin\varphi \\ u_2\cos\varphi \\ \vartheta\sin\varphi \end{pmatrix}, \tag{7.81}$$

and the ring moves as an absolutely rigid body along the horizontal axis.

If we seek vibrations that are symmetrical relative to the vertical axis, that is

$$\xi(\varphi) = \begin{pmatrix} Q^1(\varphi) \\ Q^2(\varphi) \\ M(\varphi) \end{pmatrix} = \begin{pmatrix} Q^1 \sin i\varphi \\ Q^2 \cos i\varphi \\ M \sin i\varphi \end{pmatrix},$$
$$\eta(\varphi) = \begin{pmatrix} u_1(\varphi) \\ u_2(\varphi) \\ \vartheta(\varphi) \end{pmatrix} = \begin{pmatrix} u_1 \cos i\varphi \\ u_2 \sin i\varphi \\ \vartheta \cos i\varphi \end{pmatrix}, \quad i = 1, 2, \ldots, \tag{7.82}$$

then we get the same expressions (7.80) for the natural frequencies of oscillations λ_i.

7.3 Free Vibrations of a Circular Plate

It was shown in Chap. 2 that the equations of the transverse vibrations of a plate can lead to the operator form (2.15) with respect to generalized forces ξ and displacements η. These equations, in turn, can be written as an equation in terms of displacements η

$$N\eta + R\ddot{\eta} = f.$$

In the case of the free harmonic vibrations of a circular plate of uniform thickness with frequency λ, we have the equation for displacement amplitude

$$N\,\eta - \lambda^2 R\,\eta = 0, \quad N = DB^{-1}D^*,$$

or

$$\Delta\Delta u - \lambda^2 \beta\,\rho\,h\,u = 0. \tag{7.83}$$

In the general case in the polar coordinate system, $u(r,\varphi)$ is the transverse deflection of the plate; r the variable radius, $0 \le r \le R$; φ the angular coordinate, $0 \le \varphi \le 2\pi$; $\beta = \frac{12(1-\nu^2)}{E h^3}$ the bending compliance of the plate; $\rho\, h$ the mass per unit area of the plate; E the modulus of elasticity and ν is Poisson's ratio

$$\Delta u = \frac{1}{r}\frac{\partial}{\partial r}\left(r\frac{\partial u}{\partial r}\right) + \frac{1}{r^2}\frac{\partial^2 u}{\partial \varphi^2}. \tag{7.84}$$

Let us consider the case of the axially symmetric vibrations of a circular plate of uniform thickness. We represent the amplitude of oscillations by $u = u(r)$. The Laplacian is

$$\Delta u = \frac{1}{r}\frac{\mathrm{d}}{\mathrm{d}r}\left(r\frac{\mathrm{d}u}{\mathrm{d}r}\right), \tag{7.85}$$

and λ, from (7.83), is the frequency of axially symmetric vibrations of the plate. The expressions for the angle ϑ of rotation of the cross-section, the bending moment M_r and shear force Q_r per unit length are of the form

$$\vartheta = -\frac{\mathrm{d}u}{\mathrm{d}r}, \quad M_r = -\frac{1}{\beta}\left(\frac{\mathrm{d}^2u}{\mathrm{d}r^2} + \frac{\nu}{r}\frac{\mathrm{d}u}{\mathrm{d}r}\right), \quad Q_r = -\frac{1}{\beta}\frac{\mathrm{d}}{\mathrm{d}r}\Delta u. \tag{7.86}$$

There are four possible types of simple homogeneous boundary conditions depending on the fastening conditions of the plate at $r = R$

(1) Simply supported edge:

$$u = 0, \quad M_r = 0. \tag{7.87}$$

(2) Free edge:

$$Q_r = 0, \quad M_r = 0. \tag{7.88}$$

(3) Fixed edge:

$$u = 0, \quad \vartheta = 0. \tag{7.89}$$

(4) Sliding bearing:

$$\vartheta = 0, \quad Q_r = 0. \tag{7.90}$$

By entering the dimensionless coordinate $\zeta = \frac{r}{R}$, we can write the equation for $u = u(\zeta)$ as

$$\Delta\Delta u - \lambda^2\beta\rho h R^4 u = 0 \tag{7.91}$$

or

$$\Delta\Delta u - \alpha^4 u = 0, \tag{7.92}$$

where

$$\Delta = \frac{1}{\zeta}\frac{\mathrm{d}}{\mathrm{d}\zeta}\left(\zeta\frac{\mathrm{d}}{\mathrm{d}\zeta}\right), \tag{7.93}$$

$$\alpha^4 = \lambda^2\beta\rho h R^4, \tag{7.94}$$

and $\alpha = \alpha_n, n = 1, 2, \ldots$ are the eigenvalues of the problem (i.e. those values of the parameter for which there exist nontrivial solutions).

The boundary conditions (7.87)–(7.90) at $\zeta = 1$ take the form

(1) Simply supported edge:

$$u = 0, \frac{d^2 u}{d\zeta^2} + \frac{\nu}{\zeta}\frac{du}{d\zeta} = 0. \tag{7.95}$$

(2) Free edge:

$$\frac{d^2 u}{d\zeta^2} + \frac{\nu}{\zeta}\frac{du}{d\zeta} = 0, \quad \frac{d}{d\zeta}\Delta u = 0. \tag{7.96}$$

(3) Built-in edge:

$$u = 0, \frac{du}{d\zeta} = 0. \tag{7.97}$$

(4) Sliding support:

$$\frac{du}{d\zeta} = 0, \quad \frac{d}{d\zeta}\Delta u = 0. \tag{7.98}$$

The differential equation (7.92) can be written in the symbolic form

$$(\Delta - \alpha^2)(\Delta + \alpha^2)u = 0. \tag{7.99}$$

The solution of each equation

$$\Delta u + \alpha^2 u = 0, \tag{7.100}$$

$$\Delta u - \alpha^2 u = 0 \tag{7.101}$$

will be the solution of Eq. (7.99). The solution of Eq. (7.100)

$$u(\zeta) = A\, J_0(\alpha\zeta) + B\, Y_0(\alpha\zeta).$$

The solution of Eq. (7.101)

$$u(\zeta) = C\, I_0(\alpha\zeta) + D\, K_0(\alpha\zeta),$$

where $J_0(\alpha\zeta)$, $I_0(\alpha\zeta)$ are the cylindrical and modified Bessel functions of the first kind; and $Y_0(\alpha\zeta)$, $K_0(\alpha\zeta)$ are the Weber and Macdonald functions. Since

$$|Y_0(\alpha\zeta)| \to \infty, \quad |K_0(\alpha\zeta)| \to \infty \quad \text{at} \quad |\zeta| \to 0,$$

then from the condition of boundedness of the solution at the point $\zeta = 0$, it follows that $B = 0$, $D = 0$. Hence, the solution of Eq. (7.92) can be written in the form

$$u(\zeta) = A\,J_0(\alpha\zeta) + C\,I_0(\alpha\zeta). \tag{7.102}$$

Having satisfied the boundary conditions (7.87)–(7.90) at $\zeta = 1$, we obtain the frequency equations for these cases

(1) Simply supported edge:

$$\begin{aligned} &A\,J_0(\alpha) + C\,I_0(\alpha) = 0, \\ &A(-\alpha\,J_0(\alpha) + (1-\nu)J_1(\alpha)) + C(\alpha\,I_0(\alpha) - (1-\nu)I_1(\alpha)) = 0. \end{aligned} \tag{7.103}$$

The frequency equation

$$\frac{I_1(\alpha)}{I_0(\alpha)} + \frac{J_1(\alpha)}{J_0(\alpha)} - \frac{2\alpha}{1-\nu} = 0. \tag{7.104}$$

has $\alpha = \alpha_n, n = 1, 2, \ldots$ as its solution.

(2) Free edge:

$$\begin{aligned} &\alpha = 0, \\ &A\,J_1(\alpha) + C\,I_1(\alpha) = 0, \\ &A(-\alpha\,J_0(\alpha) + (1-\nu)J_1(\alpha)) + C(\alpha\,I_0(\alpha) - (1-\nu)I_1(\alpha)) = 0. \end{aligned} \tag{7.105}$$

The frequency equation is

$$\begin{aligned} &\alpha = 0, \\ &\frac{I_0(\alpha)}{I_1(\alpha)} + \frac{J_0(\alpha)}{J_1(\alpha)} - \frac{2(1-\nu)}{\alpha} = 0. \end{aligned} \tag{7.106}$$

(3) Built-in edge:

$$\begin{aligned} A J_0(\alpha) + C I_0(\alpha) &= 0, \\ -A\alpha J_1(\alpha) + C\,\alpha I_1(\alpha) &= 0. \end{aligned} \tag{7.107}$$

The frequency equation is

$$\frac{I_1(\alpha)}{I_0(\alpha)} + \frac{J_1(\alpha)}{J_0(\alpha)} = 0. \tag{7.108}$$

4) Sliding support:

$$\begin{aligned} \alpha &= 0, \\ -AJ_1(\alpha) + CI_1(\alpha) &= 0, \\ AJ_1(\alpha) + CI_1(\alpha) &= 0, \end{aligned} \tag{7.109}$$

The frequency equation is

$$\alpha = 0, \quad J_1(\alpha) = 0. \tag{7.110}$$

Therefore, the normalized eigenfunctions $u_n(\zeta)$ corresponding to the eigenvalues $\alpha_n n = 1, 2, \ldots$, are of the form

$$u_n(\zeta) = \frac{1}{N_n}(J_0(\alpha_n\zeta) - C_n I_0(\alpha_n\zeta)), \quad C_n = -\frac{C}{A}. \tag{7.111}$$

If $\alpha_n = 0$ is an eigenvalue, then there is a form $u_1 = \frac{1}{N_1}$ corresponding to it. Here $u_n(\zeta)$ represent forms that are orthonormal with the weight ζ

$$\begin{aligned} &\int_0^1 \zeta u_n(\zeta) u_m(\zeta) \mathrm{d}\zeta = \delta_{nm}, \quad n, m = 1, 2, \ldots \\ &N_n^2 = \int_0^1 \zeta (J_0(\alpha_n\zeta) - C_n I_0(\alpha_n\zeta))^2 d\zeta. \end{aligned} \tag{7.112}$$

Let us compute N_n for all boundary conditions. We partition (7.112) into three integrals

$$\begin{aligned} N_n^2 = &\int_0^1 \zeta J_0^2(\alpha_n\zeta) \mathrm{d}\zeta - 2C_n \int_0^1 \zeta J_0(\alpha_n\zeta) I_0(\alpha_n\zeta) \mathrm{d}\zeta \\ &+ C_n^2 \int_0^1 \zeta I_0^2(\alpha_n\zeta) \mathrm{d}\zeta. \end{aligned}$$

We know that $J_0(\alpha_n\zeta)$ is the solution to the Bessel equation

$$(\zeta J_0')' + \alpha_n^2 \zeta J_0 = 0, \tag{7.113}$$

where the prime denotes a derivative with respect to the dimensionless coordinate ζ. Let us multiply (7.113) termwise by $(\zeta J_0')$ and integrate it over ζ in the range (0, 1)

$$\int_0^1 (\zeta J_0')' (\zeta J_0') \mathrm{d}\zeta + \alpha_n^2 \int_0^1 \zeta^2 J_0 J_0' \mathrm{d}\zeta = 0.$$

Transforming the integrals and denoting

$$N_{n1} = \int_0^1 \zeta J_0^2(\alpha_n \zeta) \mathrm{d}\zeta,$$

we obtain

$$N_{n1} = \frac{1}{2}\left(J_0^2(\alpha_n) + J_1^2(\alpha_n)\right). \tag{7.114}$$

Similarly, the Bessel function $I_0(\alpha_n \zeta)$ is a solution of the equation

$$(\zeta I_0')' - \alpha_n^2 \zeta I_0 = 0. \tag{7.115}$$

Carrying out the same computations as above and denoting

$$N_{n3} = \int_0^1 \zeta I_0^2(\alpha_n \zeta) \mathrm{d}\zeta,$$

we get

$$N_{n3} = \frac{1}{2}\left(I_0^2(\alpha_n) - I_1^2(\alpha_n)\right). \tag{7.116}$$

Calculating the integral

$$N_{n2} = \int_0^1 \zeta J_0(\alpha_n \zeta) I_0(\alpha_n \zeta) \mathrm{d}\zeta \tag{7.117}$$

and taking Eq. (7.113) and the properties of the Bessel functions into account gives

$$N_{n2} = \frac{1}{2\,\alpha_n}(J_0(\alpha_n)I_1(\alpha_n) + J_1(\alpha_n)I_0(\alpha_n)). \tag{7.118}$$

This gives the definitive solution

$$\begin{aligned} N_n^2 = N_{n1} - 2C_n N_{n2} + C_n^2 N_{n3} &= \frac{1}{2}\left(J_0^2(\alpha_n) + J_1^2(\alpha_n)\right) \\ &- \frac{C_n}{\alpha_n}(J_0(\alpha_n)I_1(\alpha_n) + J_1(\alpha_n)I_0(\alpha_n)) \\ &+ \frac{C_n^2}{2}\left(I_0^2(\alpha_n) - I_1^2(\alpha_n)\right), \end{aligned} \tag{7.119}$$

where the coefficients C_n for the different conditions of fastening the edge $\zeta = 1$ have the form

(1) simply supported or built-in edge:

$$C_n = \frac{J_0(\alpha_n)}{I_0(\alpha_n)}.$$

(2) free edge:

$$C_n = \frac{J_1(\alpha_n)}{I_1(\alpha_n)}.$$

(3) sliding support:

$$C_n = 0.$$

In the case of nonaxisymmetric vibrations of a circular plate of uniform thickness, we have for the function $u(\zeta, \varphi)$

$$\Delta u = \frac{1}{\zeta}\frac{\partial}{\partial \zeta}\left(\zeta \frac{\partial u}{\partial \zeta}\right) + \frac{1}{\zeta^2}\frac{\partial^2 u}{\partial \varphi^2}. \tag{7.120}$$

Let

$$u(\zeta, \varphi) = u(\zeta)\Phi(\varphi), \quad \Phi(\varphi) = \cos(n\varphi), \quad n = 1, 2, \dots \tag{7.121}$$

The case $n = 0$ of axially symmetric oscillations has been considered above. For the function $u(\zeta)$, we obtain the equation

$$\left(\frac{\mathrm{d}^2}{\mathrm{d}\zeta^2} + \frac{1}{\zeta}\frac{\mathrm{d}}{\mathrm{d}\zeta} - \frac{n^2}{\zeta^2}\right)^2 u - \alpha^4 u = 0, \quad \alpha^4 = \lambda^2 \beta \rho h R^4, \tag{7.122}$$

or two Bessel equations of the order n

$$u'' + \frac{1}{\zeta}u' + \left(\alpha^2 - \frac{n^2}{\zeta^2}\right)u = 0, \tag{7.123}$$

$$u'' + \frac{1}{\zeta}u' - \left(\alpha^2 + \frac{n^2}{\zeta^2}\right)u = 0. \tag{7.124}$$

The solution of Eq. (7.123) is

$$u(\zeta) = A\,J_n(\alpha\zeta) + B\,Y_n(\alpha\zeta),$$

and the solution of Eq. (7.124) is

$$u(\zeta) = C I_n(\alpha\zeta) + D K_n(\alpha\zeta).$$

From the condition of boundedness of the solution at $\zeta = 0$, it follows that $B = 0$, $D = 0$, and then the complete solution of Eq. (7.122) will be

$$u(\zeta) = A J_n(\alpha\zeta) + C I_n(\alpha\zeta). \tag{7.125}$$

Let us consider the different types of homogeneous boundary conditions at $\zeta = 1$ and the relevant frequency equations for these cases

(1) Simply supported edge:

$$u = 0, \quad M_r = -\frac{1}{\beta R^2}\left(u'' + \nu\left(\frac{1}{\zeta}u' - \frac{n^2}{\zeta^2}u\right)\right) = 0.$$

The frequency equation is

$$2\alpha J_n(\alpha)I_n(\alpha) - (1-\nu)(J_n(\alpha)I_{n+1}(\alpha) + J_{n+1}(\alpha)I_n(\alpha)) = 0. \tag{7.126}$$

(2) Free edge:

$$M_r = 0, \quad \tilde{Q}_r = -\frac{1}{\beta R^3}\frac{\partial}{\partial\zeta}(\tilde{\Delta}u) + \frac{n^2}{\beta R^3}(1-\nu)\frac{1}{\zeta^2}\left(u' - \frac{u}{\zeta}\right) = 0,$$

where

$$\tilde{\Delta}u = \left(u'' + \frac{1}{\zeta}u' - \frac{n^2}{\zeta^2}u\right).$$

The frequency equation in this case is

$$\begin{gathered}\alpha = 0,\\ 4n^2\alpha^2(n-1)(1-\nu)J_n(\alpha)I_n(\alpha)\\ +2\alpha^4(1-\nu)J_{n+1}(\alpha)I_{n+1}(\alpha)\\ +\left(2\alpha^3 n(n-1)(1-\alpha) - \alpha n^2(n^2-1)(1-\nu)^2 - \alpha^5\right)J_n(\alpha)I_{n+1}(\alpha)\\ -\left(2\alpha^3 n(n-1)(1-\nu) + \alpha n^2(n^2-1)(1-\nu)^2 + \alpha^5\right)J_{n+1}(\alpha)I_n(\alpha) = 0.\end{gathered} \tag{7.127}$$

(3) Built-in edge:

$$u = 0, \quad u' = 0.$$

The frequency equation is

$$J_n(\alpha)I_{n+1}(\alpha) + J_{n+1}(\alpha)I_n(\alpha) = 0. \tag{7.128}$$

(4) Sliding support:

$$u' = 0, \quad \tilde{Q}_r = 0$$

The frequency equation is

$$\begin{gathered} \alpha = 0, \\ -2\alpha n^2 J_n(\alpha)I_n(\alpha) \\ + (2\alpha^2 n + n^2(1-\nu))J_{n+1}(\alpha)I_n(\alpha) \\ + 2\alpha^3 J_{n+1}(\alpha)I_{n+1}(\alpha) \\ -(2\alpha^2 n - n^2(1-\nu))J_n(\alpha)I_{n+1}(\alpha) = 0. \end{gathered} \tag{7.129}$$

The orthonormal natural forms of the nonaxisymmetric free oscillations of a plate can be written as

$$U_{kn} = \frac{1}{N_{kn}} u_{kn}(\zeta)\cos n\varphi, \quad n, k = 1, 2, \ldots, \tag{7.130}$$

where

$$u_{kn}(\zeta) = J_n(\alpha_k\zeta) - C_k I_n(\alpha_k\zeta). \tag{7.131}$$

From the conditions of orthonormality of the forms of free oscillations with the weight ζ

$$\int_0^{2\pi}\int_0^1 \zeta\, U_{kn}(\zeta,\varphi)U_{rm}(\zeta,\varphi)\mathrm{d}\zeta\,\mathrm{d}\varphi = \begin{cases} 0, & k \neq r, \quad n \neq m, \\ 1, & k = r, \quad n = m \end{cases} \tag{7.132}$$

it follows that

$$N_{kn}^2 = \int_0^{2\pi} \cos^2 n\varphi\,\mathrm{d}\varphi \int_0^1 \zeta\, u_{kn}^2(\zeta)\mathrm{d}\zeta \tag{7.133}$$

or

$$N_{kn}^2 = \pi \int_0^1 \zeta(J_n(\alpha_k\zeta) - C_k I_n(\alpha_k\zeta))^2\mathrm{d}\zeta. \tag{7.134}$$

Performing calculations in the same way as done above, we obtain an expression for the square of the norm through the values of the Bessel functions themselves

$$\begin{aligned}\frac{N_{kn}^2}{\pi} = \frac{1}{2}\left(J_n^2(\alpha_k) + J_{n+1}^2(\alpha_k) - \frac{2n}{\alpha_k}J_n(\alpha_k)J_{n+1}(\alpha_k)\right)\\ - \frac{C_k}{\alpha_k}(I_n(\alpha_k)J_{n+1}(\alpha_k) + J_n(\alpha_k)I_{n+1}(\alpha_k))\\ + \frac{C_k^2}{2}\left(I_n^2(\alpha_k) - I_{n+1}^2(\alpha_k) - \frac{2n}{\alpha_k}I_n(\alpha_k)I_{n+1}(\alpha_k)\right),\end{aligned} \tag{7.135}$$

where the coefficients C_k for the different conditions of fastening the edge $\zeta = 1$ have the form

(1) simply supported or built-in edge:

$$C_k = \frac{J_n(\alpha_k)}{I_n(\alpha_k)},$$

(2) free edge:

$$C_k = \frac{(n(n-1)(1-\nu) - \alpha_k^2)J_n(\alpha_k) + \alpha_k(1-\nu)J_{n+1}(\alpha_k)}{(n(n-1)(1-\nu) + \alpha_k^2)I_n(\alpha_k) - \alpha_k(1-\nu)I_{n+1}(\alpha_k)}.$$

(3) sliding support:

$$C_k = \frac{n\,J_n(\alpha_k) - \alpha_k J_{n+1}(\alpha_k)}{n\,I_n(\alpha_k) + \alpha_k I_{n+1}(\alpha_k)}.$$

7.4 Vibrations of a Rectangular Plate

Let us consider the free oscillations of a thin rectangular plate of uniform thickness. Suppose the axes x and y are arranged in the middle plane of the plate and the axis z is directed vertically. We denote by $w(x, y)$ the transverse displacement of the points of the middle surface; $\beta = \frac{12(1-\nu^2)}{E\,h^3}$ the bending pliability of the plate; h the thickness of the plate; $\rho\,h$ the mass per unit area; and λ the frequency of free oscillations.

The equation for displacements describing the free transverse vibrations of a plate with frequency λ has the form

$$\frac{1}{\beta}\Delta\Delta\,w - \lambda^2\rho\,h\,w = 0$$

or

$$\frac{1}{\beta}\left(\frac{\partial^4 w}{\partial x^4}+2\frac{\partial^4 w}{\partial x^2\partial y^2}+\frac{\partial^4 w}{\partial y^4}\right)-\lambda^2\rho h w=0. \tag{7.136}$$

The equation is valid in the area

$$0\leq x\leq a,\quad 0\leq y\leq b. \tag{7.137}$$

For the convenience of calculations, we introduce the dimensionless coordinates ζ and χ that are associated (linked) with the coordinates x and y by the relations

$$\zeta=\frac{x}{a},\quad \chi=\frac{y}{a}. \tag{7.138}$$

Moreover, let us suppose

$$\mu_1=\frac{a}{b},\quad \mu=\frac{1}{\mu_1}=\frac{b}{a}. \tag{7.139}$$

Then, the area Ω occupied by the plate will be

$$0\leq\zeta\leq 1,\quad 0\leq\chi\leq\mu, \tag{7.140}$$

while Eq. (7.136) will be rewritten as

$$\left(\frac{\partial^4 w}{\partial \zeta^4}+2\frac{\partial^4 w}{\partial \zeta^2\partial \chi^2}+\frac{\partial^4 w}{\partial \chi^4}\right)-\alpha^4 w=0, \tag{7.141}$$

where

$$\alpha^4=\lambda^2\rho h\beta a^4. \tag{7.142}$$

Example Let us look at a plate pivotally supported at the edges. The boundary conditions corresponding to vanishing deflection and bending moment have the form

$$\begin{aligned}\zeta=0,\quad \zeta=1:\quad w=0,\quad \frac{\partial^2 w}{\partial \zeta^2}=0.\\ \chi=0,\quad \chi=\mu:\quad w=0,\quad \frac{\partial^2 w}{\partial \chi^2}=0.\end{aligned} \tag{7.143}$$

The natural forms of vibrations w_{mn} satisfying the conditions (7.143) and Eq. (7.141) will be recorded as follows

$$w_{mn}(\zeta, \chi) = A_{mn} \sin m\pi\zeta \sin n\pi \mu_1 \chi. \tag{7.144}$$

After substituting (7.144) into Eq. (7.141), we obtain

$$(m\pi)^4 + 2(m\pi)^2(n\,\pi\mu_1)^2 + (n\,\pi\mu_1)^4 - \alpha^4 = 0,$$

for the determination of the natural frequencies of oscillations, where

$$\alpha^2 = \pi^2(m^2 + n^2\mu_1^2),$$

and the frequencies of the free oscillations of a pivotally supported plate can be determined with the formula

$$\lambda_{mn} = \frac{\pi^2}{a^2}(m^2 + n^2\mu_1^2)\frac{1}{\sqrt{\beta\,\rho\,h}}. \tag{7.145}$$

By varying the values m and n, we can obtain the different nodal lines that divide the plate into separate parts for the corresponding natural forms of oscillations [1]. It is not difficult to verify that the forms of free oscillations determined by formula (7.144) satisfy the conditions of orthonormality

$$\int\limits_{\Omega} \rho\, h w_{mn} w_{ki} \mathrm{d}\Omega = \begin{cases} 0, & m \neq k, \quad n \neq i, \\ 1, & m = k, \quad n = i, \end{cases} \tag{7.146}$$

hence

$$A_{mn} = 2\sqrt{\frac{a}{\rho h b}}. \tag{7.147}$$

Let us consider the forced harmonic vibrations of a rectangular plate, whose sides $x = 0, x = a$ are clamped and sides S are pivotally supported. Suppose the transverse force distributed over the area load with the frequency $y = 0, y = b$ acts on the plate

$$p(x, y, t) = p(x, y)\cos(\lambda\, t + \alpha). \tag{7.148}$$

Let us write the equations of the oscillations of the plate as a pair of equations that relate to the moments and displacement as

$$\begin{aligned} &D\xi - \lambda^2 R\,\eta - f = 0, \\ &D^*\eta - B\,\xi = 0, \end{aligned} \tag{7.149}$$

where $f = p(x, y), \eta = w(x, y), \xi = (M_1, M_2, M_{12})$.

The boundary conditions recorded on the forces and displacements in the relative coordinates are as follows

(1) the conditions of sealing

$$\text{at}\, \zeta = 0, \quad \zeta = 1, \quad 0 \le \chi \le \mu : \quad w = 0, \quad \frac{\partial w}{\partial \zeta} = 0, \tag{7.150}$$

(2) the conditions of simple support

$$\text{at}\, \chi = 0, \quad \chi = \mu, \quad 0 \le \zeta \le 1 : \quad w = 0, \quad M_2 = 0. \tag{7.151}$$

The solution of the original (initial) problem can be represented as

$$\eta(\zeta,\chi) = \tilde{\eta}(\zeta,\chi), \quad \xi(\zeta,\chi) = \tilde{\xi}(\zeta,\chi) + \widehat{\xi}(\zeta), \tag{7.152}$$

where $\tilde{\eta}$, $\tilde{\xi}$ are the displacements and moments satisfying the conditions of the simply supported edges of the plate; and $\widehat{\xi}$ is an additive to the bending moment M_1, which will allow us to satisfy the conditions of absence of the rotation angle on the sides of the plate $\zeta = 0, \zeta = 1, 0 \le \chi \le \mu$.

The functions $\tilde{\eta}$, $\tilde{\xi}$ in turn can be represented as the series expansions

$$\tilde{\eta} = \sum_m \sum_n \beta_{mn} \tilde{\eta}_{mn}(\zeta,\chi), \quad \tilde{\xi} = \sum_m \sum_n \alpha_{mn} \tilde{\xi}_{mn}(\zeta,\chi) \tag{7.153}$$

in the forms of the free oscillations of the plate $\tilde{\eta}_{mn}$, $\tilde{\xi}_{mn}$ satisfying the equations

$$\begin{aligned} D\,\tilde{\xi}_{mn} - \tilde{\lambda}^2_{mn} R\,\tilde{\eta}_{mn} &= 0, \\ D^*\tilde{\eta}_{mn} - B\,\tilde{\xi}_{mn} &= 0 \end{aligned} \tag{7.154}$$

and the conditions of simple support of the plate edges

$$\begin{aligned} &\zeta = 0, \quad \zeta = 1, \quad 0 \le \chi \le \mu : \quad \tilde{w}_{mn} = 0, \quad \tilde{M}_{1_{mn}} = 0, \\ &\chi = 0, \quad \chi = \mu, \quad 0 \le \zeta \le 1 : \quad \tilde{w}_{mn} = 0, \quad \tilde{M}_{2_{mn}} = 0. \end{aligned} \tag{7.155}$$

These functions are of the form

$$\tilde{\eta}_{mn} = \tilde{w}_{mn} = A_{mn} \sin m\pi\zeta \, \sin n\pi\,\mu_1\chi, \tag{7.156}$$

$$\tilde{\xi}_{mn} = \frac{A_{mn}}{\beta a^2} \begin{pmatrix} ((m\pi)^2 + \nu (n\pi\mu_1)^2) \sin m\pi\zeta \, \sin n\pi\mu_1\chi \\ ((n\pi\mu_1)^2 + \nu (m\pi)^2) \sin m\pi\zeta \, \sin n\pi\mu_1\chi \\ -(1-\nu) m n \pi^2 \mu_1 \cos m\pi\zeta \, \cos n\pi\mu_1\chi \end{pmatrix}, \tag{7.157}$$

$$\tilde{\lambda}_{mn} = \frac{\pi^2}{a^2}(m^2 + n^2\mu_1^2)\frac{1}{\sqrt{\beta\rho h}}, \quad A_{mn} = 2\sqrt{\frac{a}{\rho h b}} \tag{7.158}$$

and satisfy the conditions of orthonormality

$$\int_{\Omega} R\,\tilde{\eta}_{mn}\,\tilde{\eta}_{ki}\mathrm{d}\Omega = \begin{cases} 0, & m \neq k, n \neq i, \\ 1, & m = k, n = i, \end{cases} \tag{7.159}$$

$$\int_{\Omega} \tilde{\xi}_{mn} B\,\tilde{\xi}_{ki}\mathrm{d}\Omega = \begin{cases} 0, & m \neq k, n \neq i, \\ \tilde{\lambda}^2_{mn}, & m = k, n = i. \end{cases} \tag{7.160}$$

In turn, the function $\widehat{\xi}\,(\zeta)$ can be represented as

$$\widehat{\xi}(\zeta) = \sum_{j=1}^{2} a_j \widehat{\xi}_j(\zeta) \tag{7.161}$$

and the following can be assumed

$$\widehat{\xi}_1(\zeta) = \begin{pmatrix} 1-\zeta \\ 0 \\ 0 \end{pmatrix}, \quad \widehat{\xi}_2(\zeta) = \begin{pmatrix} \zeta \\ 0 \\ 0 \end{pmatrix}. \tag{7.162}$$

With such a choice of coordinate functions, the boundary conditions at zero deflections are automatically fulfilled

$$\begin{aligned} &w\,(0,\chi) = 0, \quad w\,(1,\chi) = 0, \quad 0 \leq \chi \leq \mu, \\ &w\,(\zeta,0) = 0, \quad w\,(\zeta,\mu) = 0, \quad 0 \leq \zeta \leq 1 \end{aligned}$$

as are the conditions of equating the bending moment M_2 to zero

$$M_2(\zeta,0) = 0, \quad M_2(\zeta, \quad \mu) = 0, \quad 0 \leq \zeta \leq 1.$$

To determine the unknown coefficients of the expansions (7.153) and (7.161), we use a system of projection conditions

$$\begin{aligned} &\int_{\Omega} (D\xi - \lambda^2 R\,\eta - f)\tilde{\eta}_{mn}\mathrm{d}\Omega = 0, \\ &\int_{\Omega} \tilde{\xi}_{mn}(D^*\eta - B\,\xi)d\Omega = 0, \quad m,n = 1,2,\ldots \end{aligned} \tag{7.163}$$

Converting the separate terms in (7.163) by taking the properties of adjointness of the operators D and D^* in the sense of Lagrange, Eqs. (7.154), conditions (7.159), (7.160) and the boundary conditions (7.150), (7.151) and (7.155) into account, we obtain

$$\int_{\Omega} D\xi\, \tilde{\eta}_{mn} \mathrm{d}\Omega = \int_{\Omega} \xi D^* \tilde{\eta}_{mn} \mathrm{d}\Omega + \int_{0}^{\mu} \left(M_1 \frac{\partial \tilde{w}_{mn}}{\partial \zeta} \right)_{\zeta=0}^{\zeta=1} \mathrm{d}\chi$$
$$= \int_{\Omega} \xi B\, \tilde{\xi}_{mn} \mathrm{d}\Omega + \int_{0}^{\mu} \left(M_1 \frac{\partial \tilde{w}_{mn}}{\partial \zeta} \right)_{\zeta=0}^{\zeta=1} \mathrm{d}\chi$$
$$= \tilde{\lambda}_{mn}^2 \alpha_{mn} + \sum_{j=1}^{2} a_j (B_{jmn} + C_{jmn}),$$

where

$$B_{jmn} = \int_{\Omega} \hat{\xi}_j B\, \tilde{\xi}_{mn} \mathrm{d}\Omega,$$
$$C_{1mn} = \frac{m A_{mn}}{n\, \mu_1} ((-1)^n - 1), C_{2mn} = (-1)^{m+1} C_{1mn} \tag{7.164}$$

and

$$\int_{\Omega} \tilde{\xi}_{mn} D^* \eta \mathrm{d}\Omega = \int_{\Omega} D\, \tilde{\xi}_{mn} \eta \mathrm{d}\Omega = \tilde{\lambda}_{mn}^2 \int_{\Omega} R\, \tilde{\eta}_{mn} \eta \mathrm{d}\Omega = \tilde{\lambda}_{mn}^2 \beta_{mn}.$$

As a result of converting the projection conditions (7.163), we arrive at a system of algebraic equations

$$\tilde{\lambda}_{mn}^2 \alpha_{mn} - \lambda^2 \beta_{mn} + \sum_{j=1}^{2} a_j (B_{jmn} + C_{jmn}) - \tilde{f}_{mn} = 0,$$
$$\tilde{\lambda}_{mn}^2 \beta_{mn} - \tilde{\lambda}_{mn}^2 \alpha_{mn} - \sum_{j=1}^{2} a_j B_{jmn} = 0, \quad m, n = 1, 2, \ldots, \tag{7.165}$$

where $\tilde{f}_{mn} = \int_{\Omega} f\, \tilde{\eta}_{mn} \mathrm{d}\Omega$.

Equations (7.165) allow us to express all coefficients α_{mn}, β_{mn} through the coefficients a_j, $j = 1, 2$

$$\beta_{mn} = \frac{1}{\tilde{\lambda}_{mn}^2 - \lambda^2} \left(\tilde{f}_{mn} - \sum_{j=1}^{2} a_j C_{jmn} \right),$$
$$\alpha_{mn} = \beta_{mn} - \frac{1}{\tilde{\lambda}_{mn}^2} \sum_{j=1}^{2} a_j B_{jmn}, \quad m, n = 1, 2, \ldots. \tag{7.166}$$

To fulfill the conditions of equality to zero for the rotation angle of a plate with clamped edges, we begin by multiplying the second equation (7.149) sequentially by the functions $\widehat{\xi}_j(\zeta), j = 1, 2$, and then integrate them over the area Ω

$$\int_{\Omega} \widehat{\xi}_j D^* \eta \, \mathrm{d}\Omega - \int_{\Omega} \widehat{\xi}_j B \, \xi \, \mathrm{d}\Omega = 0, \quad j = 1, 2. \tag{7.167}$$

The transformation (7.167) gives

$$\int_{\Omega} \widehat{\xi}_j D^* \eta \, \mathrm{d}\Omega = -\int_0^{\mu} \left(\widehat{M}_{1_j} \frac{\partial w}{\partial \zeta} \right)_{\zeta=0}^{\zeta=1} \mathrm{d}\chi = \begin{cases} \int_0^{\mu} \left. \frac{\partial w}{\partial \zeta} \right|_{\zeta=0} \mathrm{d}\chi, & j = 1, \\ -\int_0^{\mu} \left. \frac{\partial w}{\partial \zeta} \right|_{\zeta=1} \mathrm{d}\chi, & j = 2. \end{cases}$$

$$\int_{\Omega} \widehat{\xi}_j B \, \xi \, \mathrm{d}\Omega = \sum_m \sum_n \alpha_{mn} B_{jmn} + \sum_{r=1}^{2} a_r \widehat{B}_{jr},$$

where $\widehat{B}_{jr} = \int_{\Omega} \widehat{\xi}_j B \, \widehat{\xi}_r \, \mathrm{d}\Omega, j, r = 1, 2$.

Since it is necessary to fulfill the conditions on the edges of the plate at $0 \le \chi \le \mu$

$$\left. \frac{\partial w}{\partial \zeta} \right|_{\zeta=0} = 0, \quad \left. \frac{\partial w}{\partial \zeta} \right|_{\zeta=1} = 0,$$

then, as a result, we get two additional equations

$$\sum_m \sum_n \alpha_{mn} B_{jmn} + \sum_{r=1}^{2} a_r \widehat{B}_{jr} = 0, \quad j = 1, 2. \tag{7.168}$$

Substituting α_{mn} from (7.166), expressed by a_j, we obtain a system of two algebraic equations for the coefficients a_1 and a_2

$$A \, z = b, \tag{7.169}$$

where $z = (a_1, a_2), \quad A = A\,(\lambda^2)$. Having determined z from (7.169), and taking (7.166), (7.161) and (7.153) into account, we can find a solution to the original problem of the forced harmonic oscillations of a plate that has two edges simply supported and the other two clamped.

7.5 Free Oscillations of a Spherical Shell

Let us consider a closed shell of constant thickness h, the middle surface of which is the sphere of radius R. The point position on the sphere can be defined using coordinates (β, φ). A family of parallels ($\beta = \text{const}$) and meridians ($\varphi = \text{const}$) form an orthogonal grid of coordinate curves of the middle surface and the meridian arc is measured toward the pole $\beta = 0$

$$s = R(\pi - \beta), \quad 0 \leq \beta \leq \pi.$$

We introduce unit tangent vectors to the coordinate lines $\underline{\tau}_1$, $\underline{\tau}_2$ and the unit normal vector $\underline{n}$.

Under the action of the static load, which changes along the meridian, the shell bends. The stress state can be characterized by a set of tensile forces and bending moments (T_1, T_2, N, M_1, M_2).

Along with forces and moments, we introduce the components of deformation $(\varepsilon_1, \varepsilon_2, \kappa_1, \kappa_2)$ and the elastic displacement vector $\underline{V}(\beta)$. The vector of axisymmetric displacement can be recorded by components u, w

$$\underline{V}(\beta) = u\,\underline{\tau}_1 + w\,\underline{n}.$$

The equilibrium equations of a shell under the action of an axisymmetric load, recorded on displacements, are [1]

$$\begin{aligned} L^{11}(u) + L^{12}(w) &= F_1(\beta), \\ L^{21}(u) + L^{22}(w) &= F_2(\beta), \end{aligned} \tag{7.170}$$

where

$$\begin{aligned} L^{11}(u) &= \frac{\mathrm{d}}{\mathrm{d}\beta}\frac{1}{\sin\beta}\frac{\mathrm{d}}{\mathrm{d}\beta}(\sin\beta \cdot u) + (1-\mu)u, \\ L^{21}(u) &= \frac{1+\mu}{\sin\beta}\frac{\mathrm{d}}{\mathrm{d}\beta}(\sin\beta \cdot u) - \frac{h^2}{12R^2}\Delta\left(\frac{1}{\sin\beta}\frac{\mathrm{d}}{\mathrm{d}\beta}(\sin\beta \cdot u)\right), \\ L^{12}(w) &= -(1+\mu)w + \frac{h^2}{12R^2}\frac{\mathrm{d}}{\mathrm{d}\beta}(\Delta w), \\ L^{22}(w) &= -2(1+w) - \frac{h^2}{12R^2}(\Delta\Delta w + (1-\mu)\Delta w). \end{aligned} \tag{7.171}$$

The operators are

$$\Delta = \frac{1}{\sin\beta}\frac{\mathrm{d}}{\mathrm{d}\beta}\sin\beta\frac{\mathrm{d}}{\mathrm{d}\beta}, \tag{7.172}$$

$$F_1(\beta) = \frac{1-\mu^2}{Eh}R^2 q_1(\beta), \quad F_2(\beta) = \frac{1-\mu^2}{Eh}R^2 q_2(\beta), \tag{7.173}$$

where $q_1(\beta), q_2(\beta)$ are components of the external specific (unit) load $\underline{q}(\beta)$ in the meridian and the normal directions

$$\underline{q}(\beta) = q_1(\beta)\underline{\tau}_1 + q_2(\beta)\underline{n};$$

E is the elastic modulus; and μ is the lateral compression ratio.

The equations of free vibrations can be obtained by replacing the external forces in Eqs. (7.170) by the forces of inertia

$$\underline{q} \to -0h\frac{\partial^2 \underline{V}}{\partial t^2} = -\rho h\left(\frac{\partial^2 u}{\partial t^2}\underline{\tau}_1 + \frac{\partial^2 w}{\partial t^2}\underline{n}\right) \tag{7.174}$$

or replacing q_1, q_2, respectively, by $-\rho h\frac{\partial^2 u}{\partial t^2}, -\rho h\frac{\partial^2 w}{\partial t^2}$.

The equations can be solved under the boundary conditions

$$|\underline{V}(0,t)| < \infty, \quad |\underline{V}(\pi,t)| < \infty. \tag{7.175}$$

Presenting the displacement vector $\underline{V}(\beta, t)$ as

$$\underline{V}(\beta,t) = \underline{V}(\beta)\sin(\lambda t + \alpha), \tag{7.176}$$

we obtain a system of equations for determining the forms of free oscillations $\underline{V}(\beta) = u\underline{n}$

$$\begin{aligned} L^{11}(u) + L^{12}(w) + \alpha^2 u &= 0, \\ L^{21}(u) + L^{22}(w) + \alpha^2 w &= 0 \end{aligned} \tag{7.177}$$

under the conditions

$$|u, w| \begin{matrix} \beta = 0, \\ \beta = \pi \end{matrix} < \infty, \tag{7.178}$$

where λ is the frequency of free oscillations

$$\alpha^2 = \frac{1-\mu^2}{E}\rho R^2 \lambda^2. \tag{7.179}$$

The solution of Eqs. (7.177) with conditions (7.178) can be expressed through the spherical harmonics (functions) [2]

$$
\begin{aligned}
w_n(\beta) &= P_n(\cos\beta), \quad n = 0, 1, 2, \ldots, \\
u_n(\beta) &= \eta_n P_n^1(\cos\beta), \quad n = 1, 2, \ldots,
\end{aligned}
\tag{7.180}
$$

where $P_n(\cos\beta)$, $P_n^1(\cos\beta)$ are, respectively, the Legendre polynomial and the associated Legendre polynomial that satisfy the equations

$$
\begin{aligned}
&\Delta\varphi(\beta) + n(n+1)\varphi(\beta) = 0, \quad \varphi_n = P_n(\cos\beta), \\
&\Delta\,\psi(\beta) + \left(n(n+1) - \frac{1}{\sin^2\beta}\right)\psi(\beta) = 0, \quad \psi_n = P_n^1(\cos\beta),
\end{aligned}
\tag{7.181}
$$

as a result of which $P_n^1(\cos\beta) = \frac{\mathrm{d}P_n(\cos\beta)}{\mathrm{d}\beta}$ and the functions $P_n(\cos\beta)$ and $P_n^1(\cos\beta)$ are orthogonal

$$
\begin{aligned}
&\int_0^\pi P_n(\cos\beta)P_k(\cos\beta)\sin\beta\,\mathrm{d}\beta = \frac{2}{2n+1}\delta_{nk}, \\
&\int_0^\pi P_n^1(\cos\beta)P_k^1(\cos\beta)\sin\beta\,\mathrm{d}\beta = \frac{2}{2n+1}\frac{n(n+1)}{(n-1)}\delta_{nk},
\end{aligned}
\tag{7.182}
$$

and the parameter η_n equals

$$
\eta_n = \frac{1 + \mu + \frac{1-\mu^2}{k^4}n(n+1)}{n(n+1) - 1 + \mu - \alpha_n^2}, \quad k^4 = 12(1-\mu^2)\frac{R^2}{h^2}.
\tag{7.183}
$$

The unknowns α^2 can be determined from the equation

$$
\begin{aligned}
&\alpha^4 + \left(\frac{h^2}{12R^2}p_n^2 + p_n - 1 - 3\mu\right)\alpha^2 \\
&\quad - \frac{h^2}{12R^2}p_n^2(p_n - 4) - (1-\mu^2)(p_n + 2) = 0,
\end{aligned}
\tag{7.184}
$$

which can be derived from system (7.177) using relations (7.181) and $p_n = n(n+1)$, $n = 0, 1, 2, \ldots$.

The forms of free oscillations $\underline{V}_k = u_k\underline{\tau}_1 + w_k\underline{n}$, $k = 1, 2, \ldots$, $\underline{V}_0 = \underline{n}$, $(u_0 = 0, w_0 = 1)$ are orthogonal functions; that is

$$(\underline{V}_i, \underline{V}_j) = \int_0^{\pi} \sin\beta\, \underline{V}_i \cdot \underline{V}_j \mathrm{d}\beta = \int_{-1}^{1} \underline{V}_i \cdot \underline{V}_j \mathrm{d}(\cos\beta) = 0, \quad i \neq j. \qquad (7.185)$$

From (7.179) we get $\lambda = \lambda_k$ as the frequencies of free oscillations.

References

1. Green, A.E. (1962). On the linear theory of thin elastic shells. *The Royal Society*, 266 pp.
2. Lebedev, N.N. (1776). Special functions and their application. *Dover Books in Mathematics*, 336 pp.
3. Timoshenko, S. P., & Young, D. H. (1955). *Vibration problems in engineering* (Vol. IX, p. 468). New York: Van Nostrand.

Chapter 8
Nonlinear Periodic Oscillations

8.1 Periodic Oscillations of a Thin Rod, Lying on a Nonlinear Elastic Foundation

The equation of bending vibrations of a thin rod lying on a nonlinear elastic foundation can be written as

$$\begin{aligned} & -\frac{\partial^2 M}{\partial x^2} + \rho\ddot{u} + g\dot{u} + f(u) - p = 0, \\ & -\frac{\partial^2 u}{\partial x^2} - \beta M - k = 0. \end{aligned} \tag{8.1}$$

These equations differ from Eq. (1.51) by the presence of a member $f(u)$, which takes the action of a nonlinear elastic foundation into account. Moreover, we introduce the forces of viscous friction $g\dot{u}$ in consideration of this. The notation is exactly the same as specified in Chap. 1. Considering the coefficient of bending elasticity $\beta = \beta(x) \geq (\beta(x))_{\min} > 0$ and assuming the external distributed moment as $k = k(x, t) = 0$, let us move on to a single equation with respect to displacement

$$A(u) = 0, \tag{8.2}$$

where

$$A(u) = Nu + \rho\ddot{u} + g\dot{u} + f(u) - p, \quad N = \frac{\partial^2}{\partial x^2}\left(\frac{1}{\beta}\frac{\partial^2}{\partial x^2}\right), \tag{8.3}$$

$$u = u(x, t), \quad 0 \leq x \leq l, \quad 0 \leq t \leq T, \tag{8.4}$$

l is the length of a rod; and T is the oscillation period.

The periodicity conditions should be added to Eq. (8.1)

V. Fridman, *Theory of Elastic Oscillations*, Foundations of Engineering Mechanics, DOI 10.1007/978-981-10-4786-2_8

$$u(x,0) = u(x,T), \quad \dot{u}(x,0) = \dot{u}(x,T). \tag{8.5}$$

We assume that

$$\rho(x) \geq (\rho(x))_{\min} > 0, \quad g(x) \geq (g(x))_{\min} > 0, \quad 0 \leq x \leq l.$$

We further assume that the function $f(u) = f(x,t)$ is a periodic function of time t with the periodic function $u\,(x,t)$ in a domain, which will be designated below when presenting a method for solving equations of the type (8.2). For simplicity, let us also assume that its mean (static) component is equal to zero.

Furthermore, the first and second derivatives of the function $A(u)$ with respect to the function u will be required. We introduce $\Delta = \Delta(x,t)$ as the variation of function $u = u(x,t)$. We consider this function is also twice differentiable with respect to coordinate and time, and satisfies both the boundary conditions and periodicity conditions of the form (8.5). Let us use general formulas for the differential and derivative of the nth order [7]

$$\begin{aligned} \mathrm{d}^n A(u,\Delta) &= \left(\frac{\mathrm{d}^n}{\mathrm{d}\alpha^n} A(u+\alpha\Delta)\right)\Big|_{\alpha=0}, \\ \mathrm{d}^n A(u,\Delta) &= A^{(n)}(u)\Delta^n. \end{aligned} \tag{8.6}$$

In accordance with formulas (8.6), instead of u we substitute $u + \alpha\Delta$ into formula (8.3), differentiate the obtained result with respect to α and then, setting the value α equal to 0, we get

$$A'(u) = \frac{\mathrm{d}A(u)}{du} = N + \rho\frac{\mathrm{d}^2}{dt^2} + g\frac{\mathrm{d}}{\mathrm{d}t} + \frac{\mathrm{d}f(u)}{du} \tag{8.7}$$

and

$$A''(u) = \frac{\mathrm{d}^2 A(u)}{\mathrm{d}u^2} = \frac{\mathrm{d}^2 f(u)}{\mathrm{d}u^2}. \tag{8.8}$$

Let us compose the functional as

$$(A'u, \dot{u}) = \int_0^T \int_0^l \left(Nu + \rho\ddot{u} + g\dot{u} + \frac{\mathrm{d}f(u)}{\mathrm{d}u}u\right)\dot{u}\,\mathrm{d}x\mathrm{d}t. \tag{8.9}$$

The functions $u = u(x,t)$ and $\dot{u} = \dot{u}(x,t)$ satisfy the periodicity conditions (8.5). Therefore, there will be

$$\int_0^T N u\dot{u}\mathrm{d}t = N\int_{u(x,0)}^{u(x,T)} u\mathrm{d}u = N\frac{u^2}{2}\Big|_{u(x,0)}^{u(x,T)} = 0,$$

$$\rho\int_0^T \ddot{u}\dot{u}\mathrm{d}t = \rho\int_{\dot{u}(x,0)}^{\dot{u}(x,T)} \dot{u}\mathrm{d}\dot{u} = \frac{\rho\dot{u}^2}{2}\Big|_{u(x,0)}^{u(x,T)}.$$

Moreover, as already mentioned, we consider that the function $f = f(u)$ is also a periodic function of time at the periodic function $u = u(x,t)$ and that the mean (static) component of this function is equal to zero. Then

$$\int_0^T \frac{\mathrm{d}f(u)}{\mathrm{d}u}u\dot{u}\mathrm{d}t = \int_{u(x,0)}^{u(x,T)} \frac{\mathrm{d}f(u)}{\mathrm{d}u}u\mathrm{d}u = -\int_{u(x,0)}^{u(x,T)} f(u)\mathrm{d}u = 0.$$

As a result, we obtain

$$(A'u,\dot{u}) \geq g_{\min}\int_0^T\int_0^l \dot{u}^2\mathrm{d}x\mathrm{d}t = g_{\min}||\dot{u}||^2, \quad g_{\min} = (g(x))_{\min}. \tag{8.10}$$

Let us consider an oscillatory process. For simplicity we also suppose that the static component of the displacement is equal to zero and represent a vibrating (variable) component of displacement in the form of a Fourier series expansion

$$u(x,t) = \sum_{n=-\infty}^{n=\infty} a^n(x)e^{in\lambda t}, \quad n \neq 0$$

and

$$\dot{u}(x,t) = \lambda\sum_{n=-\infty}^{n=\infty} ina^n(x)e^{in\lambda t}, \quad n \neq 0,$$

where $\lambda = \frac{2\pi}{T}$.

Hence, it follows that $|\dot{u}|| \geq \lambda||u||$, where $||u||^2 = \int_0^T\int_0^l u^2\mathrm{d}x\mathrm{d}t$.

On the other hand, according to the Schwarz inequality $(A'u,\dot{u}) \leq ||A'u||\,||\dot{u}||$ and taking formula (8.10) into account, we finally come to the important estimate

$$||A'u|| \geq g_{\min}\lambda||u||, \quad g_{\min} > 0. \tag{8.11}$$

This inequality means that there exists an inverse bounded linear operator

$$G = (A')^{-1}, \quad ||G|| \leq \frac{1}{g_{\min}\lambda}. \tag{8.12}$$

The function $f(u)$ is considered to be sufficiently smooth so as to satisfy

$$||A''(u)|| = \left\| \frac{d^2 f(u)}{du^2} \right\| \leq \left\| \frac{d^2 f(u)}{du^2} \right\|_{\max} < \infty. \tag{8.13}$$

Estimates of the form (8.12) and (8.13) are used in the theorem on the convergence of the iterative Newton-Kantorovich method.

8.2 The Newton-Kantorovich Method for Solving Nonlinear Operator Equations

Let us consider a nonlinear operator equation

$$A(u) = 0, \tag{8.14}$$

which is the same as Eq. (8.2). The equation is called "functional" since both the argument and value of the operation $A(u)$ are functions representing those of time and coordinates in the case considered above. Let us suppose the domain of definition and domain of values of the operator A are Hilbert spaces to which the notion of scalar products of elements of the space has been introduced. In the problem considered in Sect. 8.1 the elements of a Hilbert space are the functions $u\,(x, t)$, and the scalar product of two such functions can be calculated as the integral of their product over the coordinates and time.

The squared norm of the Hilbert space element is defined as the scalar product of this element with itself. Other properties of the operation $A(u)$ will be discussed below.

Let us denote the exact solution of Eq. (8.14) by $u = u_*$. Suppose also that $u = u_0$ is a function close to the solution (one of the solutions) of this equation in a certain sense. The iterative Newton-Kantorovich method of solution of a nonlinear equation (8.14) is based on approaching linear approximation of the nonlinear operation $A(u)$ in a bounded domain containing u_0. Let us represent

$$A(u) \approx A(u_0) + A'(u_0)(u - u_0)$$

approximately and put

$$A(u_0)+A'(u_0)(u-u_0)=0 \tag{8.15}$$

instead of (8.14).

Considering u_0 as the initial zero approximation of the iterative process, we will take the solution of the linear equation (8.15) for its first approximation u_1 to be

$$A(u_0)+A'(u_0)(u_1-u_0)=0.$$

Similarly, for the nth step there will be

$$A(u_n)+A'(u_n)(u_{n+1}-u_n)=0 \tag{8.16}$$

or

$$u_{n+1}=u_n-G(u_n)A(u_n), \quad G(u_n)=(A'(u_n))^{-1}. \tag{8.17}$$

This is the theorem of Kantorovich [6].

Let us suppose the following conditions are fulfilled:

(1) For the element u_0 (an initial approximation) the operation $A'(u_0)$ is the inverse of it (i.e. $G_0=(A'(u_0))^{-1}$), and an estimate of its norm can be given as

$$||G_0|| \le B_0. \tag{8.18}$$

(2) The element u_0 approximately satisfies Eq. (8.14) such that

$$||G_0A(u_0)|| \le \eta_0. \tag{8.19}$$

(3) The second derivative $A''(u)$ is bounded

$$||A''(u)|| \le K \tag{8.20}$$

in the domain defined by the inequality

$$||u-u_0|| \le 2\eta_0. \tag{8.21}$$

(4) The constants B_0, η_0, K satisfy the inequality

$$h_0=B_0\eta_0K \le \frac{1}{2}. \tag{8.22}$$

Then Eq. (8.14) has the solution u_*, which is close to u_0 in the domain defined with the same inequality (8.21), and the iterative process (8.16) or (8.17) converges to solution of Eq. (8.14).

Note that, in the problem of the bending vibrations of a thin rod on a nonlinear elastic foundation, we have proved that conditions (8.18), (8.20) are satisfied under both certain properties of the elastic foundation and the mandatory presence of viscous friction forces. Moreover, conditions (8.19), (8.21) and (8.22) have to be fulfilled for the proof of existence and convergence.

8.3 Iterative Gradient Method for Solving Operator Equations

The iterative gradient method can also be used to solve the nonlinear operator equation

$$A(u) = 0, \tag{8.23}$$

where the derivatives of the operators A' and A'' must satisfy conditions (8.12), (8.13). However, the operator A' is also required to be bounded

$$||A'|| \leq ||A'||_{\max} < \infty. \tag{8.24}$$

The domain of definition of the operator A', where this inequality needs to be fulfilled, is specified in the theorem of the convergence of the iterative gradient process. The integral equations and systems of algebraic equations satisfy the requirement (8.24). The iterative gradient method can also be used to solve differential equations approximately as long as, with the help of the projection procedure, the differential equation can first be approximately replaced by the system of algebraic equations.

The successive approximations can be calculated by the formula

$$u_{n+1} = u_n + \varepsilon_n z_n, \tag{8.25}$$

where the "step length" ε_n is a real number; and the gradient of the functional is z_n. Hence

$$l(u) = \frac{1}{2}(A(u), A(u)). \tag{8.26}$$

The gradient of the functional $l(u)$ at the point u can be determined by the formula

$$\frac{\mathrm{d}l(u+\varepsilon z)}{\mathrm{d}\varepsilon}|_{\varepsilon=0} = (z, \operatorname{grad} l(u)). \tag{8.27}$$

Let us choose an element of the domain of definition of the functional (8.26) as z, an element that imparts the maximum value to the derivative (8.27) subject to the normalization condition

$$||z|| = ||\operatorname{grad} l(u)||. \tag{8.28}$$

Clearly, such an element is precisely the gradient itself

$$z = \operatorname{grad} l(u);$$

so, we put

$$z_n = \operatorname{grad} l(u_n) \tag{8.29}$$

in formula (8.25).

Let us calculate $\operatorname{grad} l(u)$ for the functional (8.26). The derivative (8.27) of this functional takes the form

$$\begin{aligned}\frac{1}{2}\left(\frac{\mathrm{d}}{\mathrm{d}\varepsilon}(A(u+\varepsilon z), A(u+\varepsilon z))\right)|_{\varepsilon=0} &= \frac{1}{2}((A'z, A(u)) + (A(u), A'z)) \\ &= (z, (A')^* A(u)),\end{aligned}$$

and its maximum value subject to the normalization condition (8.28) is obviously attained at

$$z = (A')^* A(u).$$

Accordingly, we put

$$z_n = (A')^* A(u_n). \tag{8.30}$$

in formula (8.25).

Let us choose the coefficient ε_n from the condition of error norm minimum at each step of the iterative process

$$||u_{n+1} - u_*||^2 = \min. \tag{8.31}$$

The squared error norm is

$$||u_{n+1} - u_*||^2 = ||u_n + \varepsilon_n z_n - u_*||^2 = ||u_n - u_*||^2 + 2\varepsilon_n(u_n - u_*, z_n) + \varepsilon_n^2(z_n, z_n).$$

Obviously, the first derivative of this quantity with respect to ε_n vanishes at

$$\varepsilon_n = -\frac{(u_n - u_*, z_n)}{(z_n, z_n)}, \tag{8.32}$$

while the second derivative, equal to $2(z_n, z_n)$, is positive. Consequently, the error minimum condition (8.31) can be carried out at ε_n of (8.32).

Let us substitute z_n from (8.30) into formula (8.32)

$$\varepsilon_n = -\frac{(u_n - u_*, (A')^* A(u_n))}{((A')^* A(u_n), (A')^* A(u_n))} = -\frac{(A'u_n - A'u_*, A(u_n))}{((A')^* A(u_n), (A')^* A(u_n))}.$$

However, if the element u_n is close to the solution u_*, then there will be

$$A'u_n - A'u_* \approx A(u_n), \tag{8.33}$$

and the formula for the coefficient ε_n, which allows the minimum condition of the error norm (8.31) to be approximately fulfilled, takes the form

$$\varepsilon_n = -\frac{(A(u_n), A(u_n))}{((A')^* A(u_n), (A')^* A(u_n))} = -\frac{\|A(u_n)\|^2}{\|(A')^* A(u_n)\|^2}. \tag{8.34}$$

When the operator of Eq. (8.23) is linear, the approximate equality (8.33) becomes exact

$$A'u_n - A'u_* = A(u_n), \tag{8.35}$$

and the error minimum condition (8.31) is also satisfied accurately. Combining formulas (8.25), (8.30), (8.34), we obtain the iterative algorithm (iterative "method with minimal errors")

$$u_{n+1} = u_n - \frac{\|A(u_n)\|^2}{\|(A')^* A(u_n)\|^2} (A')^* A(u_n). \tag{8.36}$$

Convergence of the iterative procedure (8.36) is investigated in greater depth in [3, 5].

When the iterative algorithm (8.36) is applied to solve a linear operator equation, convergence of the process can be proved making the most general assumption of existence of a solution, even though it is not the only one [4]. Which of the possible solutions turns out to be the limit of successive approximations can be determined by the choice of the initial approximation.

An iterative formula of the gradient method, as in the Newton-Kantorovich method, presupposes knowledge of the first derivative A'. The advantage of the gradient method rests in use by the adjoint operator $(A')^*$ of the multiplication operation in an iterative algorithm, which is much simpler than calculation of the inverse operation $G = (A')^{-1}$, as in the Newton-Kantorovich method. The disadvantage of the gradient method is its slow convergence. However, the increasing performance of computers will promote wider application of the gradient method.

8.4 Nonlinear Vibrations, Close to Periodic

Close-to-periodic linear oscillations of a lumped mass attached to a spring were examined in Sect. 5.6. The approach presented there can also be extended to the case of nonlinear oscillations. We use the same designations, assumptions and some algorithms given in that section.

The equation of nonlinear oscillations of a lumped mass on a spring will be written as

$$\frac{d^2u}{dt^2} + \gamma\left(\tau, \frac{du}{dt}\right) + c(\tau, u) = f(\tau, t), \tag{8.37}$$

where the term $\gamma\left(\tau, \frac{du}{dt}\right)$ takes a nonlinear frictional force acting on the oscillating mass into account, while the term $c(\tau, u)$ represents the nonlinear elastic force of the spring. All terms in the equation can be divided by the mass amount.

We still assume that the external force $f(\tau, t)$ at a fixed slow time τ is a periodic function of a fast time t with the frequency $\omega(\tau)$ and period $T(\tau) = \frac{2\pi}{\omega(\tau)}$. Putting the arguments of the functions in Eq. (8.37) in brackets indicates the nonlinear dependence of these functions of the arguments.

Taking the variable $\theta(t)$ from formula (5.41) we get $\frac{d}{dt} = \frac{d}{d\theta}\omega$. Here Eq. (8.37) can be replaced by the system of equations

$$\begin{aligned} &\frac{dv}{d\theta}\omega(\tau) + \gamma(\tau, v) + c(\tau, u) - f(\tau, \theta) = 0, \\ &\frac{du}{d\theta}\omega(\tau) - v = 0. \end{aligned} \tag{8.38}$$

An approximate solution to this system of equations at $\tau \geq 0$, as in Sect. 5.6, can be sought in the form

$$u(\tau, \theta) = \sum_{m=-M}^{M} u^m(\tau)e^{im\theta}, \quad v(\tau, \theta) = \sum_{m=-M}^{M} v^m(\tau)e^{im\theta}. \tag{8.39}$$

We give the projection conditions as

$$\begin{aligned}&\frac{1}{2\pi}\int_0^{2\pi} e^{-is\theta}\left(\frac{dv}{d\theta}\omega+\gamma(\tau,v)+c(\tau,u)-f(\tau,\theta)\right)d\theta=0,\\&\frac{1}{2\pi}\int_0^{2\pi} e^{-is\theta}\left(\frac{du}{d\theta}\omega-v\right)d\theta=0,\quad s=-M,\ldots,M.\end{aligned}\tag{8.40}$$

We also rewrite formulas (5.48) from Sect. 5.6 as

$$\begin{aligned}\frac{dv}{d\theta}&=\sum_{m=-M}^{M}\left(\frac{dv^m}{d\theta}+imv^m\right)e^{im\theta},\\\frac{du}{d\theta}&=\sum_{m=-M}^{M}\left(\frac{du^m}{d\theta}+imu^m\right)e^{im\theta}.\end{aligned}\tag{8.41}$$

Substituting the series (8.39) into the projection conditions (8.40) and using formulas (8.41), we obtain equations that can determine the coefficients of the series (8.39)

$$\begin{aligned}\frac{dv^s(\tau)}{d\theta}&=-isv^s(\tau)-\Lambda^s(u^m(\tau),v^m(\tau)),\\\frac{du^s(\tau)}{d\theta}&=-isu^s(\tau)+\frac{1}{\omega(\tau)}v^s(\tau),\quad m,s=-M,\ldots,M,\end{aligned}\tag{8.42}$$

$$\Lambda^s(u^m(\tau),v^m(\tau))=\frac{1}{2\pi\omega}\int_0^{2\pi}e^{-is\theta}(\gamma(v(\tau,\theta))+c(u(\tau,\theta))-f(\tau,\theta))d\theta,\tag{8.43}$$

where the functions $v(\tau,\theta)$ and $u(\tau,\theta)$ are represented by the series (8.39). When calculating the integral (8.43), the slow time τ can be regarded as a parameter.

Let us now suppose that when $\tau=0$ there is a purely periodic motion at $\gamma(0,v)$, $c(0,u), f(0,\theta)$ with a constant frequency $\omega\,(0)$. Entering these parameter values for external force and oscillation frequency in formulas (8.42), (8.43) and using the Runge-Kutta method, we get a movement which at the limit tends to the periodic. Representing the periodic motion obtained as a series

$$u(0,\theta)=\sum_{m=-M}^{M}u^m(0)e^{im\theta},\quad v(0,\theta)=\sum_{m=-M}^{M}v^m(0)e^{im\theta},\tag{8.44}$$

we are able to find the values $u^m(0)$ and $v^m(0)$, $m=-M,\ldots,M$, that should be added to the system of Eq. (8.42) as the initial conditions. Hence, the initial value problem (8.39), (8.42) can again be solved by the Runge-Kutta method, but only for the slowly varying parameters of the oscillating system.

The projection method for solving the problems of vibrations that are close to periodic, as outlined in Chap. 5 and earlier in this chapter, is an alternative to the asymptotic methods that introduce the concepts of fast and slow time. However, the solution using the projection method can be represented as a series expansion in a small parameter, which is usually limited to one or two approximations [1, 2, 8].

References

1. Blekhman, I. I. (1968). The small parameter method: 50 years of mechanics in the U.S.S.R. *Science, 1*, 157–165 (in Russian).
2. Bogolyubov, N. N., & Mitropol'skii, Y. A. (1961). *Asymptotic methods in the theory of nonlinear oscillations* (537 pp). New York: Cordon Science Publishers.
3. Fridman, V. M. (1961). The iterative process with minimal errors for nonlinear operator equations. *Reports of the Academy of Sciences of the U.S.S.R. (Proceedings of the Academy of Sciences), 139*(5), 1063–1066 (in Russian).
4. Fridman, V. M. (1962). On the convergence of methods of the steepest descent type. *Successes of Mathematical Sciences (Russian Mathematical Surveys), XVII*(3), (195), 201–204 (in Russian).
5. Fridman, V. M., & Pittel, B. G. (1969). On the convergence of the iteration process with minimal errors. *Proceedings of the Leningrad Polytechnic Institute, Mechanics and Control Processes, 307*, 182–188 (in Russian).
6. Kantorovich, L. V. (1948). Functional analysis and applied mathematics. *Successes of Mathematical Sciences (Russian Mathematical Surveys), 3*(6), 89–185 (in Russian).
7. Lusternik, L. A., & Sobolev, V. J. (1968). *Elements of functional analysis* (411 pp). Gordon and Breach.
8. Reimers, N. A., & Khodzhaev, K. Sh. (1978). Averaging of quasilinear systems with many fast variables. *Differential Equations, 14*(8), 1388–1399 (in Russian).

Part II
Some Applied Problems

Chapter 9
Determination of Elastic Damping Characteristics of Slide Bearings

In practice, the vibrations of elastic bodies can occur in a variety of media be they liquid, air, electromagnetic. The following chapter considers the vibrations of a rotary shaft on bearings lubricated with liquid. This chapter is devoted to determination of the elastic damping properties of such bearings. Thus, we use some information from hydrodynamic theory.

9.1 The Hydrodynamic Theory of Lubrication

The hydrodynamic theory of viscous fluid largely retains the concepts of deformable solid body dynamics. There are two basic differences. First, Hooke's law of elasticity is replaced by Newton's law of viscosity. Second, points of the medium are not linked to points in the fixed coordinate system, as in the linear elasticity theory, but move relative to this system.

Let us start by rewriting the equations of the dynamics of a three-dimensional solid (2.15) by introducing the velocity of the material point as a new variable

$$v = \dot{\eta}, \quad v = \underline{v} \tag{9.1}$$

and differentiating the second Eq. (2.15) with respect to time. We obtain

$$\begin{aligned} &D\xi + R\dot{v} - f = 0, \\ &D^* v - (B\xi)^{\dot{}} - \dot{e} = 0, \end{aligned} \tag{9.2}$$

where, yet again, $\xi = \underline{\underline{\tau}}$ is the stress tensor; $R\dot{v} = \rho\dot{\underline{v}}$ are inertial forces; $f = \underline{p}$ is the vector of volumetric forces; $e = \underline{\underline{e}}$ is the strain tensor from external causes; $D\xi = -div\,\underline{\underline{\tau}}$; and $D^* v = def\,\underline{v}$. Moreover

V. Fridman, *Theory of Elastic Oscillations*, Foundations of Engineering Mechanics, DOI 10.1007/978-981-10-4786-2_9

$$(B\xi)^{\cdot} = \dot{\underline{\underline{\Theta}}} + \dot{\underline{\underline{\Gamma}}}, \tag{9.3}$$

where $\underline{\underline{\Theta}}$ is the spherical and $\underline{\underline{\Gamma}}$ the deviatoric parts of the strain tensor.

Moving from equations of the dynamics of an elastic solid to those of the hydrodynamics of viscous fluid [2], we assume for the spherical part of the strain and stress tensors that the law of elasticity persists

$$\dot{\underline{\underline{\Theta}}} = k\dot{\underline{\underline{\sigma}}}, \tag{9.4}$$

and for the deviatoric part, instead of the law of elasticity (2.11), the law of viscosity persists

$$\dot{\underline{\underline{\Gamma}}} = \frac{1}{2\mu}\underline{\underline{S}}, \tag{9.5}$$

where μ is the dynamic viscosity coefficient. Assuming in (9.2) that $\dot{e} = 0$, we can write the equations of motion of viscous liquid as

$$\begin{aligned} &-div\,\underline{\underline{\tau}} + \rho\,\dot{\underline{v}} - \underline{p} = 0,\\ &def\,\underline{v} - \left(k\,\dot{\underline{\underline{\sigma}}} + \frac{1}{2\mu}\,\underline{\underline{S}}\right) = 0, \end{aligned} \tag{9.6}$$

in which

$$\dot{\underline{v}} = \frac{\mathrm{d}\underline{v}}{\mathrm{d}t} = \frac{\partial \underline{v}}{\partial t} + (\underline{v}\cdot\nabla)\underline{v}. \tag{9.7}$$

The first term in (9.7) is local acceleration of fluid at points fixed relative to a stationary coordinate system. The second term is convective acceleration due to the fact that points of liquid are not attached to a fixed coordinate system, but are moving in a nonuniform velocity field.

An important application of liquids with increased viscosity is the use of oil as a lubricant to reduce friction between moving machine parts.

In the hydrodynamic theory of lubrication the following assumptions are commonly used:

1. Fluid is considered incompressible; that is, it is assumed that

$$k = 0, \quad \dot{\underline{\underline{\Theta}}} = 0. \tag{9.8}$$

2. The inertial properties of grease are ignored; that is, there is a reliance that

$$\rho\dot{\underline{v}} = 0. \tag{9.9}$$

3. Volumetric forces in the oil layer can also be ignored

$$\underline{p} = 0. \tag{9.10}$$

Making these assumptions, Eqs. (9.6) for a viscous incompressible fluid take on the form

$$\begin{aligned} &\operatorname{div} \underline{\underline{\tau}} = 0, \\ &\operatorname{def} \underline{v} - \frac{1}{2\mu}(\underline{\underline{\tau}} - \underline{\underline{\sigma}}) = 0. \end{aligned} \tag{9.11}$$

The condition of fluid incompressibility (9.8) can be written as

$$\dot{\varepsilon}_x + \dot{\varepsilon}_y + \dot{\varepsilon}_z = 0$$

or

$$\frac{\partial v_x}{\partial x} + \frac{\partial v_y}{\partial y} + \frac{\partial v_z}{\partial z} = 0,$$

otherwise

$$\operatorname{div} \underline{v} = 0. \tag{9.12}$$

We apply the operator div to the second Equation (9.11)

$$\operatorname{div}(\operatorname{def} \underline{v}) - \frac{1}{2\mu} \operatorname{div} \underline{\underline{\tau}} + \frac{1}{2\mu} \operatorname{div} \underline{\underline{\sigma}} = 0. \tag{9.13}$$

Considering the first Equation (9.11), the property of fluid incompressibility (9.12) and the equality

$$\operatorname{div} \underline{\underline{\sigma}} = -\nabla p,$$

where $p = -\sigma = -\frac{1}{3}(\sigma_x + \sigma_y + \sigma_z)$ is hydrodynamic pressure at the given point of flow, we obtain the Navier-Stokes equation of viscous fluid dynamics [2]

$$\mu \, \Delta \underline{v} - \nabla p = 0, \tag{9.14}$$

which is linear with respect to velocity $\underline{v}$, since the inertial properties of the oil film have been ignored.

Equation (9.14) together with the incompressibility condition (9.12) is a system of equations in which velocity $\underline{v}$ and pressure p are relatively unknown. Boundary and initial conditions or conditions of periodicity should be added to Eqs. (9.12) and (9.14). On the part of the boundary there is distributed load, while on the other part there is movement velocity. The latter can be formulated as the condition of viscous fluid adhering to an adjacent solid body. As far as time is concerned, the conditions are imposed on velocities distributed over the entire area for which the solution is constructed.

9.2 The Theory of the Oil Wedge

The essential feature of the hydrodynamic theory of lubrication is the study of the hydrodynamics of an extremely thin layer of viscous liquid, often called the "oil film". The reason for studying the hydrodynamics of the thin layer can be explained by the capacity of this layer to:

- provide the relative sliding motion of structural elements with a low coefficient of friction;
- perceive high loads (i.e. possess a high load-bearing capacity).

The latter capacity occurs because viscous liquid sucked into a tapering convergent slot (the confuser) provides great pressure on its sidewalls. The confuser (convergent) portion of the oil film is also called the "oil wedge". The effect of oil wedge action on maintaining a rotating shaft with a horizontal axis is widely used in slide bearings (at thermal and nuclear power plants).

Let us consider the oil wedge theory proposed by the Australian scientist A. Michel [3].

Figure 9.1 shows the simplest scheme of an oil wedge in which the plane problem is considered. Along the axis perpendicular to the drawing nothing changes. The oil wedge has two lateral flat boundaries. The horizontal plane moves as indicated by the arrow at the constant velocity U and performs vertical oscillations with the velocity $V(t)$. The inclined plane is fixed. The moving surface is the outer surface of the rotating shaft journal of the bearing, and the fixed inclined plane is the inner surface of the housing liner. The rotor hub of the thrust bearing is rotary, while the thrust-bearing segments are stationary.

Fig. 9.1 Oil wedge

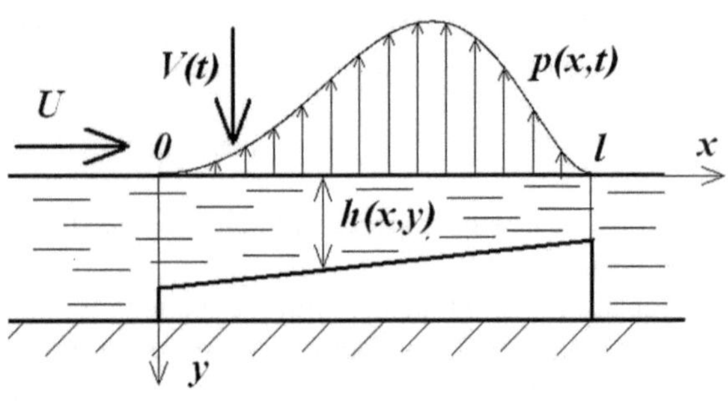

We show the axis x horizontally and the axis y vertically downwards. We put the origin of coordinates on the horizontal plane at the beginning of the wedge. The length of the wedge along the axis x is denoted by l.

The problem of the hydrodynamics of an oil wedge can be solved with the following assumptions [3]:

(1) assume the plane motion of the viscous fluid in the wedge applies to the area $0 \le x \le l,\ 0 \le y \le h(x,t)$, where $h(x,t)$ is the thickness of the oil layer;

(2) assume that by virtue of the narrowness of the wedge compared with its length $\left(\frac{h}{l} \ll 1\right)$ the pressure does not change over the layer thickness; that is

$$p = p(x,t), \qquad \frac{\partial p}{\partial y} = 0,$$

(3) assume that the change of velocities over the layer thickness is considerably greater than over its length

$$\frac{\partial^2 v_x}{\partial y^2} \gg \frac{\partial^2 v_x}{\partial x^2}, \qquad \frac{\partial^2 v_y}{\partial y^2} \gg \frac{\partial^2 v_y}{\partial x^2},$$

(4) assume that the liquid sticks to the stationary and moving surfaces of the wedge.

Making these assumptions the Navier-Stokes Eqs. (9.14) and the conditions of incompressibility of the fluid (9.12) take the form

$$\begin{aligned} &\mu \frac{\partial^2 v_x}{\partial y^2} - \frac{\partial p}{\partial x} = 0, \\ &\mu \frac{\partial^2 v_y}{\partial y^2} = 0, \\ &\frac{\partial v_x}{\partial x} + \frac{\partial v_y}{\partial y} = 0. \end{aligned} \tag{9.15}$$

Equations (9.15) are valid in the abovementioned area $0 \le x \le l,\ 0 \le y \le h(x,t)$. We assume that the pressure p (in excess of atmospheric pressure) is equal to zero at the end of the wedge at $x = 0, l$. Moreover, we take the conditions for liquid particles sticking to the surface of the wedge into account.

Hence, we get the following boundary conditions

$$\begin{array}{lll} x = 0, & 0 \le y \le h(0,t) & p = 0, \\ x = l, & 0 \le y \le h(l,t), & p = 0, \\ y = 0, & 0 \le x \le l, & v_x = U, v_y = V(t), \\ y = h(x,t), & 0 \le x \le l, & v_x = 0, v_y = 0. \end{array} \tag{9.16}$$

The reason for wedge calculation is to determine the allocation of the velocity of the fluid and the pressure therein along the coordinate x. The initial conditions or periodicity conditions imposed on the unknown quantities by time need to be added to the equations of motion of the fluid in the oil layer and the boundary conditions.

After integrating the first two Eqs. (9.15) and taking conditions (9.16) into account, we obtain expressions for the velocities

$$v_x = v_{x1} + v_{x2}, \quad v_y = V\left(1 - \frac{y}{h}\right), \tag{9.17}$$

where

$$v_{x1} = U\left(1 - \frac{y}{h}\right), \quad v_{x2} = \frac{1}{2\mu}\frac{\partial p}{\partial x}(y-h)y. \tag{9.18}$$

The third Equation (9.15) needs to be averaged over the thickness of the layer

$$\int_0^h \left(\frac{\partial v_x}{\partial x} + \frac{\partial v_y}{\partial y}\right)\mathrm{d}y = 0,$$

from which

$$\frac{\partial}{\partial x}\int_0^h v_x\,\mathrm{d}y + v_y\big|_{y=0}^{y=h} = 0$$

or

$$\frac{\partial g}{\partial x} - V = 0, \tag{9.19}$$

where $g\,(x, t)$ is the expense of fluid through the cross-section of the wedge with the coordinate x per unit time

$$g\,(x, t) = \int_0^h v_x\,\mathrm{d}y. \tag{9.20}$$

It can be seen from Eq. (9.19) that we have $g = \text{const}$ (i.e. the liquid flow rate through any section of the gap is constant in the absence of oscillations of the rotor) at $V(t) = 0$.

After substituting velocity v_x from (9.17), (9.18) into expression (9.20), we obtain

$$g(x,t) = -\frac{h^3}{12\mu}\frac{\partial p}{\partial x} + \frac{h}{2}U. \tag{9.21}$$

Excluding the flow rate $g(x,t)$ from (9.19) and (9.21), we obtain the Reynolds equation with respect to the pressure in the oil layer

$$\frac{\partial}{\partial x}\left(\frac{h^3}{12\mu}\frac{\partial p}{\partial x}\right) - q = 0, \tag{9.22}$$

where

$$q = \frac{U}{2}\frac{\partial h}{\partial x} - V. \tag{9.23}$$

We solve Eq. (9.22) under the boundary conditions

$$p(0,t) = 0, \quad p(l,t) = 0 \tag{9.24}$$

as well as the conditions on time, which are described above.

Integrating Eq. (9.22) and taking the boundary conditions (9.24) into account, we obtain

$$\frac{\partial p}{\partial x} = \lambda \int_0^x q(s,t)\,\mathrm{d}s - c, \quad c = \frac{\lambda \int_0^l \lambda\left(\int_0^x q(s,t)\,\mathrm{d}s\right)\mathrm{d}x}{\int_0^l \lambda\,\mathrm{d}x}, \quad \lambda = \frac{12\mu}{h^3} \tag{9.25}$$

and

$$p(x,t) = \int_0^x \frac{\partial p(s,t)}{\partial s}\,\mathrm{d}s. \tag{9.26}$$

If we substitute expression (9.25) for $\frac{\partial p}{\partial x}$ into the formula $\int_0^l \frac{\partial p}{\partial x}\mathrm{d}x = p(l,t) - p(0,t)$, then, as can easily be seen, we get $0 = p(l,t) - p(0,t)$. In addition, according to (9.26), $p(0,t) = 0$. Thus, both boundary conditions (9.24) are satisfied.

The total lifting force of the oil layer here can be calculated by the formula

$$P = \int_0^l p\,\mathrm{d}x = p x\big|_{x=0}^{x=l} - \int_0^l \frac{\partial p}{\partial x} x\,\mathrm{d}x = -\int_0^l \frac{\partial p}{\partial x} x\,\mathrm{d}x. \tag{9.27}$$

If we now substitute $q = q\,(s,t)$ from (9.23) into formula (9.25) and use the formula for the lifting force (9.27), then (having performed the necessary transformations), we find that

$$P = F_1(h)\,U + F_2(h)\,V, \tag{9.28}$$

where $F_1(h)$ and $F_2(h)$ are functions nonlinearly dependent on the layer thickness h.

9.3 Cylindrical Plain Bearing

Oil-lubricated cylindrical slide bearings are widely used in engineering. They serve as supports for rotating machine rotors. The lifting force of a bearing uses the effect of the oil wedge. The bearing consists of a housing, shaft journal of a rotary machine and the oil film separating them. Usually, the static component of the force acting on the neck occurs because of the weight of the rotor and is directed vertically downwards. Furthermore, variable forces can act on a bearing as the rotor rotates when the centrifugal forces of the different parts of the rotor are not balanced perfectly.

Based on the theory of the oil wedge outlined, we can determine the dynamic properties of a cylindrical plain bearing. We assume the inner surface of the bearing housing and the outer surface of the shaft journal are purely cylindrical. Let us denote the radii of these surfaces, respectively, as R_1 and R_0, $R_1 > R_0$. The radial clearance in the bearing at a concentric position of the journal relative to the housing is

$$\delta = R_1 - R_0, \tag{9.29}$$

such that $\delta \ll R_1, \delta \ll R_0$.

Under the action of vertical static load (i.e. weight), concentricity is disturbed and the journal is shifted, deviating from the vertical on one or other side according to the direction of rotation of the journal.

Let us introduce an orthogonal coordinate system x, y, where x is the horizontal axis and y the vertical axis (directed upward), with the intersection in the center of the circumference of the housing. Moreover, we denote the angular coordinate measured from the horizontal axis x by $\varphi, 0 \le \varphi \le 2\pi$, and the circumferential coordinate where $R = \frac{R_1 + R_0}{2}$ by $s = R\varphi$. Let us assume that ζ_x, ζ_y are projections on the axes x, y of the displacement of the journal center relative to the housing center. The gap between the housing and the journal can be calculated by the formula

$$h\,(s,t) = \delta - \zeta_x \cos\varphi - \zeta_y \sin\varphi, \quad \varphi = \frac{s}{R}, \tag{9.30}$$

and

$$\frac{\partial h(s,t)}{\partial s} = \frac{1}{R}\left(\zeta_x \sin\varphi - \zeta_y \cos\varphi\right),$$
$$\frac{\partial h\,(s,t)}{\partial t} = -\dot{\zeta}_x \cos\varphi - \dot{\zeta}_y \sin\varphi. \tag{9.31}$$

The circumferential velocity U at the radius R subject to rotor rotation at an angular velocity ω as well as the rate of change of gap thickness are equal to

$$U = R\omega, \quad V = \frac{\partial h}{\partial t}. \tag{9.32}$$

As a result of displacement of the journal center relative to the housing center, the clearance $h\,(s,t)$ between the inner surface of the bearing housing and the outer surface of the journal becomes, according to (9.30), variable on the circumferential coordinate s. Moreover, this clearance is found to consist of two equal parts (Fig. 9.2).

In the first part (shaded) the magnitude of the gap decreases as coordinate s increases, and in the second part, conversely, the clearance increases as s increases. The first part of the gap is called the "confuser", while the second part is referred to as the "diffuser". The confuser part of the gap is formed by the oil wedge, which provides the pressure in the oil layer and creates the lifting force of the bearing. Since negative pressure cannot be created in a liquid, we assume that pressure in the oil film is equal to zero (referring to pressure in excess of atmospheric pressure) in the diffuser part of the circumference. We denote the coordinates at the beginning and end of the confuser by $s_1 = a$ and $s_2 = b$, respectively, where $s_2 > s_1$, $s_2 - s_1 = \pi R$, $a = R\varphi_0$ and φ_0 is the angle of ascent.

Sommerfeld [3] assumed that there was no lateral spread of grease and that the active part of the bearing coincided with its convergent part. This allows us to use formulas (9.33)–(9.35) to compute the pressure in the oil wedge. Hence, we have

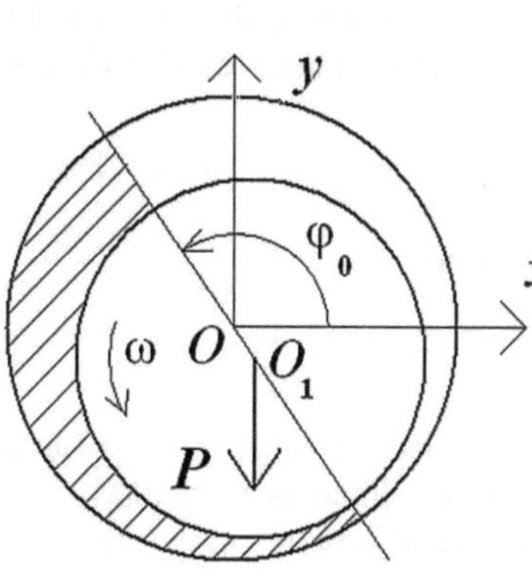

Fig. 9.2 Cylindrical bearing of Sommerfeld

$$\frac{\partial p}{\partial s} = g\int_a^s q\,(u,t)\,\mathrm{d}u - c,$$
$$c = \frac{g\int_a^b g\left(\int_a^s q(u,t)\,\mathrm{d}u\right)\mathrm{d}s}{\int_a^b g\,\mathrm{d}s}, \quad g = \frac{12\mu}{h^3}, \tag{9.33}$$

$$p\,(s,t) = \int_a^s \frac{\partial p(u,t)}{\partial u}\,du. \tag{9.34}$$

According to formulas (9.23), (9.32) and (9.33)

$$q = \frac{U}{2}\frac{\partial h}{\partial s} - V = \frac{\omega R}{2}\frac{\partial h}{\partial s} - \frac{\partial h}{\partial t} \tag{9.35}$$

at $\frac{\partial h}{\partial s}$ and $\frac{\partial h}{\partial t}$ from (9.31).

We can now calculate the projections on the axes x and y of the total lifting force of the bearing oil film (per unit axial length)

$$P_x = -\int_a^b p\cos\varphi\,\mathrm{d}s, \quad P_y = -\int_a^b p\sin\varphi\,\mathrm{d}s, \quad \varphi = \frac{s}{R}. \tag{9.36}$$

Combining formulas (9.33)–(9.35), we obtain a formula of the oil film lift force dependence on the displacement and velocity of the journal center inside the bearing housing

$$P = P\,(\zeta,\,\dot{\zeta}) = Q\,(\zeta) + S(\zeta)\dot{\zeta},$$
$$P = (P_x,\,P_y), \quad \zeta = (\zeta_x, \zeta_y). \tag{9.37}$$

If an external force $F = F(t),\ F = (F_x, F_y)$ acts on the rotor journal (again per unit axial length), then the equilibrium condition for the bearing journal, taking (9.37) into account, can be written as

$$P\,(\zeta,\,\dot{\zeta}) - F = Q\,(\zeta) + S\,(\zeta)\dot{\zeta} - F = 0.$$

Hence

$$\dot{\zeta} = S^{-1}(\zeta)\,(F - Q\,(\zeta)). \tag{9.38}$$

The Runge-Kutta method can be used for numerical solution of this equation. Such a method is particularly suitable for the external periodic force $F = F(t)$. Numerical experiments for real bearings show that the motion orbit calculated after just two or three cycles takes the form of a periodic curve, regardless of the choice of the initial approximation $\zeta(0) = \zeta_0$ (naturally, the start point selected of the orbit should be placed within the gap). Such a result can be explained by the fact that, according to Newton's law, significant forces of viscous damping act on the oil film. Moreover, the Runge-Kutta method in the appendix to Eq. (9.38) allows us to find a solution to the static problem if we put $F = F(t) = \text{const}$.

When considering small oscillations of the shaft near the point of static equilibrium, the problem can be solved as stated above. Namely, we assume that

$$\zeta(t) = \zeta_0 + z(t), \tag{9.39}$$

where $z(t)$ is the vector of small deviation of the shaft from the equilibrium point, having projections on the axes x and y.

Let us denote these displacement projections also by x and y (i.e. $z = (x, y)$) and expend the nonlinear function $P(\zeta, \dot{\zeta})$ in a Taylor series in the neighborhood of a point corresponding to the static equilibrium position $\zeta = \zeta_0$, $\dot{\zeta} = 0$, retaining only the linear terms

$$P(\zeta, \dot{\zeta}) = P(\zeta_0, 0) + \frac{\partial P}{\partial \zeta}\Big|_{\substack{\zeta=\zeta_0 \\ \dot{\zeta}=0}} \cdot z + \frac{\partial P}{\partial \dot{\zeta}}\Big|_{\substack{\zeta=\zeta_0 \\ \dot{\zeta}=0}} \cdot \dot{z}, \tag{9.40}$$

where $P(\zeta_0, 0)$ is the static reaction of the oil film.

Having expressed the force $P = (P_x, P_y)$ and displacement $z = (x, y)$ through their projections, we obtain the matrix elastic-damping characteristics of the oil film of a sliding bearing

$$\frac{\partial P}{\partial \zeta} = \begin{pmatrix} c_{xx} & c_{xy} \\ c_{yx} & c_{yy} \end{pmatrix}, \quad \frac{\partial P}{\partial \dot{\zeta}} = \begin{pmatrix} \kappa_{xx} & \kappa_{xy} \\ \kappa_{yx} & \kappa_{yy} \end{pmatrix}. \tag{9.41}$$

9.4 Cylindrical Bearing (Projection Method for Solving Pressure Distribution in the Oil Film)

In the previous section an exact solution to the problem of pressure distribution in the oil layer of a cylindrical sleeve bearing was deduced. However, such an important factor as the lateral spread of oil in a bearing of finite width was not taken into account.

An approximate solution to the problem of pressure distribution in the oil film in view of the lateral spread of lubricant can be achieved using a combination of both projection and sweep methods. Apart from the circumferential coordinate s, we keep the notation of the preceding section and introduce the axial coordinate u. To demonstrate the technique, let us first examine the static case. In statics, Eq. (9.22) can be written as follows:

$$\frac{\mathrm{d}}{\mathrm{d}s}\left(\frac{h^3}{12\mu}\frac{\mathrm{d}p}{\mathrm{d}s}\right) - q = 0, \quad q = \frac{\omega R}{2}\frac{\mathrm{d}h}{\mathrm{d}s}. \tag{9.42}$$

By taking the lateral spread of oil into account, this ordinary differential equation can be replaced by the partial differential equation

$$Ap - q = 0, \quad A = \frac{\partial}{\partial s}\left(\frac{h^3}{12\mu}\frac{\partial p}{\partial s}\right) + \frac{\partial}{\partial u}\left(\frac{h^3}{12\mu}\frac{\partial p}{\partial u}\right), \quad q = \frac{\omega R}{2}\frac{\partial h}{\partial s}, \tag{9.43}$$

where $p = p\,(s, u)$ is pressure in excess of atmospheric pressure.

We assume that Eq. (9.43) holds in the domain Ω, for which $0 \le s \le l$, $l = 2\pi R$, $0 \le u \le b$, where b is bearing width. The boundary conditions imposed on the function $p(s, u)$ are

$$p\,(0, u) = p\,(l, u), \quad \frac{\partial p}{\partial s}(0, u) = \frac{\partial p}{\partial s}(l, u), \tag{9.44}$$

$$p\,(s, 0) = p\,(s, b) = 0. \tag{9.45}$$

Let us give such values to the index m

$$m = -M, -(M-1), \ldots, (M-1), M \tag{9.46}$$

and approximately represent the inhomogeneous term in Eq. (9.43), as well as the solution of this equation, in the form of a finite harmonic series

$$q\,(s, u) = \sum_{m=-M}^{M} q^m \mathrm{e}^{im\varphi}, \quad \varphi = \frac{s}{R}, \quad q^m = q^m(u), \tag{9.47}$$

$$p\,(s, u) = \sum_{m=-M}^{M} p^m \mathrm{e}^{im\varphi}, \quad \varphi = \frac{s}{R}, \quad p^m = p^m(u). \tag{9.48}$$

The projection conditions can be written as

$$\frac{1}{2\pi}\int_0^{2\pi} \mathrm{e}^{-ik\varphi}(Ap - q)\,\mathrm{d}\varphi = 0, \quad k = -K, -(K-1), \ldots, (K-1),\ K;\ K = M. \tag{9.49}$$

Let us substitute expression (9.43) for the operator A

$$\frac{1}{2\pi}\int_0^{2\pi} \mathrm{e}^{-ik\varphi}\left(\frac{\partial}{\partial s}\left(g\frac{\partial p}{\partial s}\right)+\frac{\partial}{\partial u}\left(g\frac{\partial p}{\partial u}\right)-q\right)\mathrm{d}\varphi = 0, \quad g=\frac{h^3}{12\mu}, \tag{9.50}$$

where radial clearance in the bearing can be calculated using formula (9.30), since $h = h(s)$ is independent of time in statics. We substitute $s = R\varphi$ in Eq. (9.50) and integrate the first term by parts, taking the second boundary condition (9.44) into account. This gives us

$$\frac{1}{2\pi}\int_0^{2\pi} \mathrm{e}^{-ik\varphi} g\left(\left(\frac{ik}{R^2}\frac{\partial p}{\partial \varphi}\right)+\left(\frac{\partial^2 p}{\partial u^2}\right)\right)\mathrm{d}\varphi - q^k = 0, \quad q^k = \frac{1}{2\pi}\int_0^{2\pi} \mathrm{e}^{-ik\varphi} q \,\mathrm{d}\varphi. \tag{9.51}$$

Let us represent pressure in the form of the series (9.48) and change the order of integration and summation

$$\sum_{m=-M}^{M} g^{km}\left(\left(\frac{-km}{R^2}p^m\right)+\left(\frac{\mathrm{d}^2 p^m}{\mathrm{d}u^2}\right)\right) - q^k = 0, \tag{9.52}$$

where

$$g^{km} = \frac{1}{2\pi}\int_0^{2\pi} \mathrm{e}^{i(m-k)\varphi} g \,\mathrm{d}\varphi. \tag{9.53}$$

On the axial axis u we place N equidistant points with coordinates u_n, $n = 1, 2, \ldots, N-1, N$, where the numbers $n = 1$ and $n = N$ correspond to the edges of the bearing. Let us denote the distance between the points by Δ and apply the finite difference approximation

$$\frac{\mathrm{d}^2 p^m}{\mathrm{d}u^2} = \left(\left(p^m_{n+1} - p^m_n\right) - \left(p^m_n - p^m_{n-1}\right)\right)\frac{1}{\Delta^2}, \quad n = 2, 3, \ldots, N-2, N-1, \tag{9.54}$$

where by virtue of the boundary conditions (9.45) $p^m_1 = p^m_N = 0$.

Let us assume the indices m and k take the values (9.46) and (9.49). We now introduce the column vectors $p_n = (p^m_n)$, $q_n = (q^k_n)$ and the matrices $G_1 = \left(-\frac{km}{R^2}g^{km}\right)$, $G_2 = \left(g^{km}\right)$.

According to Eq. (9.52) and formula (9.54) we can write

$$G_1 p_n + \frac{1}{\Delta^2}\left(G_2 p_{n+1} - 2G_2 p_n + G_2 p_{n-1}\right) - q_n = 0. \tag{9.55}$$

This equation can be put into the form

$$\begin{gathered}A_{n,n-1}p_{n-1}+A_{n,n}p_n+A_{n,n+1}p_{n+1}-q_n=0,\\ A_{n,n-1}=\frac{1}{\Delta^2}G_2,\ A_{n,n}=G_1-\frac{2}{\Delta^2}G_2,\ A_{n,n+1}=\frac{1}{\Delta^2}G_2.\end{gathered} \tag{9.56}$$

The tridiagonal system (9.56) can be solved in a convenient way by using the matrix sweep method, which applies recurrence formulas of the form (4.86). Having determined the numerical coefficients p_n^m, we can represent approximately the required pressure distribution across the oil film in the form of a harmonic series of the type (9.48)

$$p(s,u_n)=\sum_{m=-M}^{M}p_n^m\mathrm{e}^{im\varphi},\quad s=\varphi R. \tag{9.57}$$

In this case fulfillment of the conditions of periodicity (9.44) is provided automatically by expansion of the solution in the harmonic series, and the conditions on the bearing edges (9.45) are essentially used in the implementation of direct and reverse sweep. After having obtained a general solution to the problem of pressure distribution in the oil layer (9.57), it is possible to specify the points around the circumference where the pressure values $s=0,\ s_1,\ldots,S-1.$ can be calculated concretely.

The number of points S can be chosen large enough for a detailed approximation of the solution. It will not have a significant impact on the overall time of calculation, since the point numbers can be substituted by a ready-made solution. The above actions mean such an operation

$$p=A^{-1}q \tag{9.58}$$

can be carried out approximately. Note that the pressure $p=p\,(s,u)$ can take both positive and negative values. Meanwhile, pressure in the oil film (in excess of atmospheric pressure) cannot be negative

$$p=p\,(s,u)\geq 0. \tag{9.59}$$

Solution of the problems (9.43), (9.44), (9.45) under the additional condition (9.59) can be done with the iterative method by using the operation A^{-1} at each step and replacing negative values of the pressure by null ones, much as was done in [1]. Consideration of this matter goes beyond the scope of the present book.

In heavily loaded bearings the positive values of pressures in the active zone calculated by formula (9.58) are considerably higher than the negative pressures in the unloaded zone. Therefore, the latter can be approximately replaced by zeros. An acceptable result can be obtained in the first approximation by using the formula

$$p=p_*\psi(p_*),\quad p_*=A^{-1}q, \tag{9.60}$$

where

$$\psi(p_*) = 1, \text{ if } p_* > 0 \quad \text{or} \quad \psi(p_*) = 0 \text{ if } p_* \leq 0.$$

Pressure distribution in the oil film of a cylindrical sliding bearing of a high-power internal combustion engine has been calculated. The number of points around the circumference was chosen as $S = 360$, while the number of points in the axial direction was chosen as $N = 31$. Overall, the pressure has been determined as $360 \cdot 31 = 11,160$ points. Computation on an average power personal computer takes a few tens of seconds. The calculated epure of pressure in the oil layer of a cylindrical sliding bearing is shown in Fig. 9.3.

Note that the proposed method for solving Eq. (9.43) can also be extended, albeit with some complications, to the case when $h = h(s, u)$—generally bounded positive differentiable with respect to s function.

The method can also be applied to the solution of dynamic problems. To do so at each instant of movement, instead of $q(s, u)$ from formula (9.42), we have to use formula (9.35) for $q(s, u, t) : q = \frac{\omega R}{2}\frac{\partial h}{\partial s} - \frac{\partial h}{\partial t}$.

The function of time $\dot{\zeta}(t) = (\dot{\zeta}_x(t), \dot{\zeta}_y(t))$, or in words the moving velocity of a rotor gudgeon relative to a bearing housing, was introduced in formula (9.31). The derivative $\frac{\partial h}{\partial t}$ can be expressed through projections of the function $\dot{\zeta}$ using formula (9.31). Moreover, the total hydrodynamic reaction of the oil layer can be calculated, similar to (9.36), by the formulas

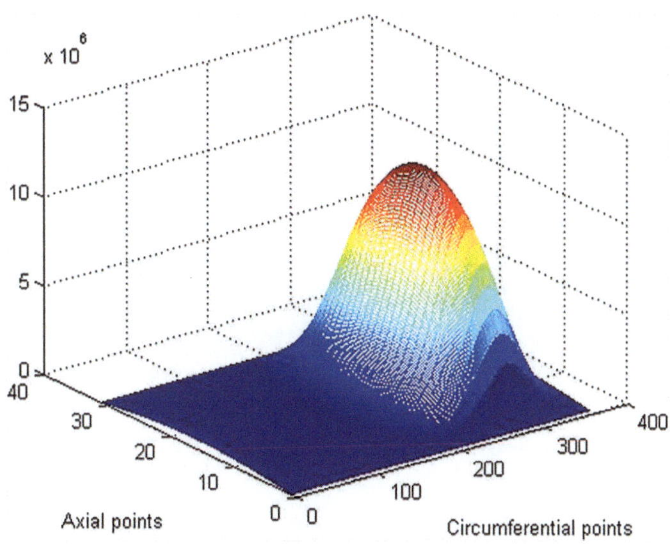

Fig. 9.3 Pressure in the oil film of a cylindrical bearing of ICE (360 · 31 points) (dimensions along the ordinate axis are given in dynes/m^2)

$$P_x = -\int_{\Omega} p\cos\varphi\, d\Omega, \quad P_y = -\int_{\Omega} p\sin\varphi\, d\Omega. \tag{9.61}$$

Thus, the relation between the movement of a rotor journal in a bearing housing and the hydrodynamic reaction of the oil layer has been established. This allows overall solution of the problem of the vibrational behavior of a bearing assembly. Calculation of the relative motion of a rotor journal and a bearing housing over a single period takes a few tens of minutes. Hence, at each instant of time the pressure values in the oil layer can be determined at all 11,160 points. Solution of the problem considered using the finite element method to solve the Reynolds equation for the large number of points indicated takes significantly more time.

References

1. Fridman, V. M., & Chernina, V. S. (1967). The iterative process for solving the finite-dimensional contact problem. *USSR Computational Mathematics and Mathematical Physics, 7,* 160–163.
2. Loitsianskii, L. G. (1995). *Mechanics of liquids and gases* (p. 840). Begell House.
3. Sommerfeld, A. K. (1934). The hydrodynamic lubrication theory. In L. S. Leibenzon (Ed.), *The hydrodynamic theory of lubrication* (pp. 361–448). Moscow: Scientific and Technical Publishing House (ONTI) (in Russian).

Chapter 10
Vibrations of Shafts, Blades and Disks

This chapter examines the oscillation problems of shafts, turbine blades and the joint vibrations of blades and disks of a steam turbine's rotor.

Shaft vibrations. Shafts are the basic carrier parts of the rotors of various types of turbomachines. The remaining parts of rotors are fastened to the shafts, and in calculations are considered as additional masses. Therefore, when discussing the vibrations of rotors, we are really talking about the vibrations of shafts. Let us consider three reasons which should be taken into account when studying these oscillations.

(1) The action of unbalanced centrifugal forces generated during rotation of a rotor.
(2) The variability of shaft parameters in two mutually perpendicular planes, such as twofold bending stiffness.
(3) The presence of support anisotropy.

10.1 Bending Vibrations of a Rotating Shaft on Pivot Supports Under Unbalanced Centrifugal Forces

Let us consider vibrations of a shaft of circular cross-section on pivot bearings. Let s be the coordinate directed along the axis of the shaft; $\rho(s)$ its mass per unit length; and $\beta(s)$ is distributed pliability at bending. We consider a perfect geometry (i.e. assume that the shaft is an axisymmetric cylinder, whose line of cross-section geometric centers (elastic line) is straight in the absence of loads (shaft is considered to be weightless)). The shaft can be regarded as a thin rod. We introduce two coordinate systems: x, y, s is stationary and u, v, s is rotary with the shaft at speed ω. The top of the coordinate system is compatible with the axis of a straight shaft. We suppose a uniform rotation such that the angle between two specified coordinate systems is considered equal to $\alpha = \omega t$.

V. Fridman, *Theory of Elastic Oscillations*, Foundations of Engineering Mechanics, DOI 10.1007/978-981-10-4786-2_10

Assume that the center of gravity of the shaft in an arbitrary cross-section in which the longitudinal coordinate s, generally speaking, does not coincide with the geometric center of this cross-section. In the coordinate system connected with the rotating shaft, we can represent the eccentricity of the cross-section gravity center of the shaft relative to its geometric center as a complex function

$$\varepsilon(s) = \varepsilon_u(s) + i\varepsilon_v(s), \tag{10.1}$$

where $\varepsilon_u(s)$ and $\varepsilon_v(s)$ are projections of eccentricity on the axes of the rotating system u, v. In a fixed coordinate system the cross-section center of gravity can be determined by the vector $\varepsilon e^{i\omega t}$.

Rotation of the shaft induces bending vibrations because of centrifugal forces. As a result the geometrical center of a cross-section with coordinate s shifts relative to the axes x and y. Let $x(s,t), y(s,t)$ be the elastic line displacement relative to the axes x and y; and $M_y(s,t), M_x(s,t)$ the bending moments relative to the axes y and x. We introduce complex deflections and moments as

$$\begin{aligned} z(s,t) &= x(s,t) + iy(s,t), \\ M(s,t) &= M_y(s,t) + iM_x(s,t). \end{aligned} \tag{10.2}$$

Then, displacement of the center of gravity of the cross-section can be determined by the vector

$$z_c = z + \varepsilon e^{i\omega t}. \tag{10.3}$$

The equations for bending vibrations of a shaft in a fixed coordinate system and boundary conditions in the complex form take the form

$$\begin{aligned} -M'' + \rho \ddot{z}_c &= 0, \\ -z'' - \beta M &= 0. \end{aligned} \tag{10.4}$$

$$M(0,t) = 0, M(l,t) = 0, z(0,t) = 0, z(l,t) = 0. \tag{10.5}$$

Equations (10.4) taking (10.3) into account can be given as

$$\begin{aligned} &- M'' + \rho \ddot{z} - pe^{i\omega t} = 0, \\ &- z'' - \beta M = 0. \end{aligned} \tag{10.6}$$

In Eqs. (10.6)

$$p = p(s) = \rho(s)\varepsilon(s)\omega^2, M = M(s,t), z = z(s,t).$$

Thus, the force $p(s,t) = \rho\,\varepsilon\,\omega^2 e^{i\omega t}$ is an external impact.

The solution of Eqs. (10.6) can be sought in the form

$$M(s,t) = M(s)e^{i\omega t},\ z(s,t) = z(s)e^{i\omega t}, \tag{10.7}$$

where $M(s), z(s)$ are the amplitude values of the moment and displacement in fixed axes or the moment and displacement in a movable coordinate system.

For $M(s)$ and $z(s)$ we have the equations

$$\begin{aligned} &-M'' - \omega^2 \rho z - p = 0,\\ &-z'' - \beta M = 0 \end{aligned} \tag{10.8}$$

and the boundary conditions

$$M(0) = 0,\ M(l) = 0,\ z(0) = 0,\ z(l) = 0. \tag{10.9}$$

Equation (10.8) can be written as the operator equation in displacements

$$N\eta - \omega^2 R\eta - f = 0, \tag{10.10}$$

where

$$\begin{aligned} &\eta = z,\ f = p,\ R = \rho,\ N\eta = DB^{-1}D^{*}\eta = \left(\frac{1}{\beta}z''\right)'',\\ &\beta = \frac{1}{EJ},\ E = E(s),\ J = J(s) = J_x(s) = J_y(s). \end{aligned} \tag{10.11}$$

The solution of (10.10) can be represented as a series

$$z(s) = \sum_{k=1}^{\infty} \frac{p_k}{\lambda_k^2 - \omega^2} z_k(s), \tag{10.12}$$

in the forms of free oscillations of the shaft $\eta_k = z_k(s)$ supported on pivot bearings, which satisfies the equation

$$N\eta_k - \lambda_k^2 R\eta_k = 0 \tag{10.13}$$

and the boundary conditions

$$\eta_k(0) = 0,\ \eta_k(l) = 0,\ \eta_k''(0) = 0,\ \eta_k''(l) = 0, \tag{10.14}$$

where $p_k = \int_0^l p z_k \mathrm{d}s = \omega^2 \int_0^l \rho\, \varepsilon z_k \mathrm{d}s$.

The solution (10.12) can be written in another form by substituting $p_k = \int_0^l p(r) z_k(r) \mathrm{d}r$ into formula (10.12)

$$z(s) = \int_0^l \Gamma(s, r) p(r) \mathrm{d}r, \tag{10.15}$$

where

$$\Gamma(s, r) = \sum_{k=1}^{\infty} \frac{z_k(s) z_k(r)}{\lambda_k^2 - \omega^2} \tag{10.16}$$

is the dynamic influence function that enables us to express deflection by the load or by using the integral operator Γ

$$z = \Gamma p. \tag{10.17}$$

As is evident in (10.7), the geometric center of the cross-section of a rotating shaft in a fixed coordinate system moves with a circular motion $z(s,t) = z(s)\mathrm{e}^{i\omega t}$ called "direct precession" of the shaft center at speed ω. From (10.3) it follows that the center of gravity of the cross-section commits direct precession as well. In a coordinate system, deflection in a rotating shaft is permanent (just like observing a still picture). This is precisely the case when variable stresses in a rotating shaft are absent.

10.2 Bending Vibrations of an Unbalanced Rotating Shaft on Isotropic Elastic Supports

Let us consider the bending vibrations of an unbalanced round shaft on isotropic elastic supports (the properties of the supports are the same in two mutually perpendicular directions). Let c_1 and c_2 be the stiffness of the left and right supports, respectively. The mixed boundary conditions

$$Q(0) = M'(0) = c_1 z(0), Q(l) = M'(l) = -c_2 z(l), \tag{10.18}$$

$$M(0) = 0, M(l) = 0. \tag{10.19}$$

should be added to Eqs. (10.8) for the bending vibrations of the shaft. Problems (10.8), (10.18), (10.19) can be solved in two ways: expanding the solution as a series in the forms of free oscillations of the shaft on elastic supports, or using the spectral properties of the shaft with simplified boundary conditions (e.g. with the conditions of simple support at the ends of the shaft). In the latter case, an algorithm that can determine the forms and frequencies of free vibrations is easier since there is no need to recalculate for different values of elasticity of the supports.

A solution to Eqs. (10.8) with conditions (10.18), (10.19) can be represented as

$$z(s) = \tilde{z}(s) + \hat{z}(s). \tag{10.20}$$

The first term in (10.20) is elastic deflection of a shaft on absolutely rigid pivot bearings, whereas the second term is displacement of a shaft as a rigid body on elastic supports.

The elastic deflection $\tilde{z}(s)$ can be represented as a series

$$\tilde{z}(s) = \sum_{k=1}^{\infty} \beta_k \tilde{z}_k(s) \tag{10.21}$$

of the forms of free oscillations $\tilde{z}_k(s)$ of a shaft on pivot bearings which satisfy the conditions of orthonormality

$$\int_0^l \rho(s)\tilde{z}_k(s)\tilde{z}_r(s)\mathrm{d}s = \delta_{kr}, \quad k, r = 1, 2, \ldots, \tag{10.22}$$

and can be determined for a shaft with a variable cross-section (e.g. using the iterative procedure outlined in Chap. 3, but under simple boundary conditions).

The deflection $\hat{z}(s)$ can be sought in the form

$$\hat{z}(s) = \sum_{j=1}^{2} b_j \hat{z}_j(s), \tag{10.23}$$

where for $\hat{z}_j(s)$ we choose the linear functions

$$\hat{z}_1(s) = 1 - \frac{s}{l}, \quad \hat{z}_2(s) = \frac{s}{l} \tag{10.24}$$

as two linearly independent solutions of the equation $N\hat{z} = 0$. Such a choice of coordinate functions shows that the coefficients b_1 and b_2 are precisely shaft displacements at its ends

$$b_1 = z(0), \quad b_2 = z(l).$$

Let us write Eqs. (10.8) in terms of displacements using the differential operator N

$$Nz - \omega^2 \rho z - p = 0. \tag{10.25}$$

Let us substitute (10.20) into Eq. (10.25) and take $N\widehat{z} = 0$ into account. Then we obtain the equation

$$N\tilde{z} - \omega^2 \rho \tilde{z} - \tilde{p} = 0, \quad \tilde{p} = p + \omega^2 \rho, \tag{10.26}$$

and homogeneous boundary conditions

$$\tilde{z}(0) = 0, \ \tilde{z}(l) = 0, \ \tilde{z}''(0) = 0, \quad \tilde{z}''(l) = 0. \tag{10.27}$$

for the function $\tilde{z}(s)$.

Thus, the problem can be reduced to solution of Eq. (10.26) with the homogeneous boundary conditions (10.27). A peculiarity lies in the fact that (10.26) includes the unknown function $\tilde{p}$ since $\widehat{z}(s)$ has not been determined yet.

The solution of (10.26), (10.27) can be constructed in the same way as (10.21), where

$$\beta_k = \frac{\tilde{p}_k}{\tilde{\lambda}_k^2 - \omega^2}, \tag{10.28}$$

and $\tilde{\lambda}_k$ are the frequencies of free bending vibrations of a shaft on pivot bearings

$$\begin{gathered} \tilde{p}_k = \int_0^l \tilde{p}\tilde{z}_k \mathrm{d}s = p_k + \omega^2 \sum_{j=1}^{2} b_j R_{jk}, \\ R_{jk} = \int_0^l \rho \widehat{z}_j \tilde{z}_k \mathrm{d}s, \quad p_k = \int_0^l p\tilde{z}_k \mathrm{d}s, \end{gathered} \tag{10.29}$$

or else

$$\beta_k = \frac{p_k}{\tilde{\lambda}_k^2 - \omega^2} + \frac{\omega^2}{\tilde{\lambda}_k^2 - \omega^2} \sum_{j=1}^{2} b_j R_{jk}, \quad k = 1, 2, \ldots. \tag{10.30}$$

As is evident from (10.30) the coefficients β_k can be expressed by the as-yet unknown coefficients b_1, b_2. The latter can be obtained from the boundary conditions.

The boundary conditions (10.18) can be performed in different ways. One way lies in the fact that the shearing force can be represented as a series in the forms of free oscillations $\tilde{Q}_k(s)$

$$Q(s) = \widehat{Q}(s) + \tilde{Q}(s) = \sum_{k=1}^{\infty} \beta_k \tilde{Q}_k(s),$$
$$\widehat{Q} = -(EJ(\widehat{z})'')' = 0, \quad \tilde{Q}_k = -(EJ\tilde{z}_k'')'. \tag{10.31}$$

Thus, the shearing force can be obtained by differentiating the deflection, which in turn can be sought iteratively.

Another way involves expressing the shearing force by integrating the load. This can be done by writing the first equation of (10.8) as

$$-Q' - \omega^2 \rho z - p = 0. \tag{10.32}$$

Multiplying (10.32) by $\widehat{z}_1(s)$ and integrating over the length of the shaft, we get

$$\int_0^l (Q' + \omega^2 \rho z + p)\widehat{z}_1 \mathrm{d}s = 0. \tag{10.33}$$

The transformation (10.33) leads to the expression

$$Q(0) = \omega^2 \left(\sum_{k=1}^{\infty} \beta_k R_{1k} + b_1 \widehat{\mathrm{R}}_{11} + b_2 \widehat{\mathrm{R}}_{21} \right) + \widehat{p}_1. \tag{10.34}$$

Similarly, multiplying (10.32) by $\widehat{z}_2(s)$ and integrating with respect to s, we get an expression for the force at $s = l$

$$-Q(l) = \omega^2 \left(\sum_{k=1}^{\infty} \beta_k R_{2k} + b_1 \widehat{\mathrm{R}}_{12} + b_2 \widehat{\mathrm{R}}_{22} \right) + \widehat{p}_2. \tag{10.35}$$

The notation

$$R_{1k} = \int_0^l \rho \widehat{z}_1 \tilde{z}_k \mathrm{d}s, \quad \widehat{\mathrm{R}}_{jr} = \int_0^l \rho \widehat{z}_j \widehat{z}_r \mathrm{d}s, \quad \widehat{p}_j = \int_0^l p \widehat{z}_j \mathrm{d}s,$$
$$j, r = 1, 2, k = 1, 2, \ldots. \tag{10.36}$$

was introduced in expressions (10.34), (10.35).

Considering the boundary conditions (10.18) for the shearing force at the ends of the shaft and expressions (10.34), (10.35), we obtain

$$c_1 b_1 - \omega^2 \left(\sum_{k=1}^{\infty} \beta_k R_{1k} + b_1 \widehat{R}_{11} + b_2 \widehat{R}_{21} \right) - \widehat{p}_1 = 0,$$
$$c_2 b_2 - \omega^2 \left(\sum_{k=1}^{\infty} \beta_k R_{2k} + b_1 \widehat{R}_{12} + b_2 \widehat{R}_{22} \right) - \widehat{p}_2 = 0. \tag{10.37}$$

After substituting the coefficients β_k, expressed by means of (10.30) through b_1 and b_2, into (10.37), we have a system of two algebraic equations

$$Ax = f \tag{10.38}$$

about $x = (b_1, b_2)$. Note that $A = A(\omega^2)$. Having found b_1 and b_2 from (10.38), coupled with (10.30), (10.20), (10.21), (10.23), we are now in a position to construct a solution to the initial problem.

Let us consider the case when the shaft is symmetrical with reference to the mid-span and its supports have the identical characteristics $c_1 = c_2 = c$. In this case we can separately solve the problems of both symmetric and antisymmetric vibrations. We do this by taking the functions

$$(\widehat{z}_1)^* = \widehat{z}_1 + \widehat{z}_2 = 1, \quad (\widehat{z}_2)^* = \widehat{z}_1 - \widehat{z}_2 = 1 - 2\frac{s}{l}$$

as $\widehat{z}_j(s), j = 1, 2$. Then, when considering symmetric vibrations in the expansions (10.21), (10.23), we should only keep symmetric forms $(\widehat{z}_1)^*, \tilde{z}_1(s), \tilde{z}_3(s), \ldots$, while when considering antisymmetric vibrations only asymmetric forms $(\widehat{z}_2)^*, \tilde{z}_2(s), \tilde{z}_4(s), \ldots$ should be kept.

Let us consider the free symmetric vibrations of a shaft on isotropic elastic supports. In the above mentioned formulas we put

$$p = 0, \quad \omega = \lambda, \quad c_1 = c_2 = c, \quad b_1 = b_2 = b,$$

where λ is the natural frequency of a shaft on elastic supports. Transforming the integrals entered in (10.37), we obtain

$$R_{1k} + R_{2k} = \int_0^l \rho \tilde{z}_k (\widehat{z}_1)^* \mathrm{d}s = R_{1k}^*,$$

$$\widehat{R}_{11} + \widehat{R}_{12} + \widehat{R}_{21} + \widehat{R}_{22} = \int_0^l \rho \mathrm{d}s = m,$$

$$\beta_k = \frac{\lambda^2}{\tilde{\lambda}_k^2 - \lambda^2} b (R_{1k} + R_{2k}) = b \frac{\lambda^2}{\tilde{\lambda}_k^2 - \lambda^2} R_{1k}^*,$$

where m is the mass of the shaft.

Combining this with Eq. (10.37), we obtain

$$b\left(2c-\lambda^2 m-\lambda^2\sum_{k=1}^{\infty}\frac{(R_{1k}^*)^2\lambda^2}{\tilde{\lambda}_k^2-\lambda^2}\right)=0,$$

where, since $b \neq 0$ are displacements of the shaft ends on elastic supports that differ from zero, the frequency equation for determining frequencies λ of the free symmetric bending vibrations of a shaft on elastic supports is as follows

$$2c-\lambda^2 m-\lambda^2\sum_{k=1}^{\infty}\frac{(R_{1k}^*)^2\lambda^2}{\tilde{\lambda}_k^2-\lambda^2}=0. \tag{10.39}$$

Hence, it can be seen that for an absolutely rigid shaft on elastic supports we get

$$\lambda^2=\frac{2c}{m}.$$

10.3 Bending Vibrations of an Unbalanced Rotating Shaft on Isotropic Elastic-Damping Supports

In the case of isotropic elastic-damping supports the equations of bending vibrations of an unbalanced rotating shaft (10.6) can be solved using the following boundary conditions

$$\begin{aligned}Q(0,t)&=c_1 z(0,t)+\kappa_1\dot{z}(0,t),\\ Q(l,t)&=-c_2 z(l,t)-\kappa_2\dot{z}(l,t),\\ M(0,t)&=0, M(l,t)=0,\end{aligned} \tag{10.40}$$

where $c_1, \kappa_1, c_2, \kappa_2$ are the stiffness and damping coefficients of the left and right supports.

Seeking a solution to the problem, just as in the case of isotropic supports, in the form of

$$M(s,t)=M(s)\mathrm{e}^{i\omega t},\quad z(s,t)=z(s)\mathrm{e}^{i\omega t},\quad Q(s,t)=Q(s)\mathrm{e}^{i\omega t}, \tag{10.41}$$

we obtain Eqs. (10.8) and the boundary conditions for the amplitude values of forces and displacements

$$\begin{gathered}Q(0)=(c_1+i\omega\,\kappa_1)z(0),\quad Q(l)=-(c_2+i\omega\,\kappa_2)z(l),\\ M(0)=0, M(l)=0.\end{gathered} \tag{10.42}$$

The method of solving the problem of vibrations of a shaft on isotropic elastic-damping supports is similar to the one outlined in the preceding section. The only difference is the appearance of an imaginary term in the boundary conditions (10.42).

10.4 Bending Vibrations of an Unbalanced Rotating Shaft on Anisotropic Elastic-Damping Supports

Let us consider the bending vibrations of an unbalanced shaft on linear anisotropic elastic-damping supports and write the expression for the shearing force projections on the support (e.g. we take the left support as $s = 0$)

$$\begin{aligned} Q_x &= c_{xx}x + c_{xy}y + \kappa_{xx}\dot{x} + \kappa_{xy}\dot{y}, \\ Q_y &= c_{yx}x + c_{yy}y + \kappa_{yx}\dot{x} + \kappa_{yy}\dot{y}. \end{aligned} \tag{10.43}$$

In particular, the elastic-damping anisotropy of the supports can be determined by the anisotropic properties of the oil-lubricated bearings, as was shown in Chap. 9
.

Proceeding to complex variables

$$z = x + iy, \quad Q = Q_x + iQ_y$$

allows us to rewrite conditions (10.43) as

$$Q = cz + d\bar{z} + \kappa\dot{z} + \nu\dot{\bar{z}}, \tag{10.44}$$

where the notation $\bar{z} = x - iy$ has been introduced, as well as

$$\begin{aligned} c &= \frac{c_{xx} + c_{yy}}{2} - i\frac{c_{xy} - c_{yx}}{2}, \quad & d &= \frac{c_{xx} - c_{yy}}{2} + i\frac{c_{xy} + c_{yx}}{2}, \\ \kappa &= \frac{\kappa_{xx} + \kappa_{yy}}{2} - i\frac{\kappa_{xy} - \kappa_{yx}}{2}, \quad & \nu &= \frac{\kappa_{xx} - \kappa_{yy}}{2} + i\frac{\kappa_{xy} + \kappa_{yx}}{2}. \end{aligned} \tag{10.45}$$

The equation for oscillations of a shaft in a stationary coordinate system using the differential operator N can be written as

$$Nz + \rho\ddot{z} - \sum_{n=-1,1} p_n e^{in\omega t} = 0, \tag{10.46}$$

where

$$p_n(s) = \begin{cases} p_1(s) = p(s), & n = 1, \\ p_{-1}(s) = 0, & n = -1. \end{cases} \tag{10.47}$$

The solution to Eq. (10.46) can be written in the form of a generalized precession of the axis of the shaft

$$z(s,t) = z_{-1}(s)\mathrm{e}^{-i\omega t} + z_1(s)\mathrm{e}^{i\omega t} = \sum_{n=-1,1} z_n(s)\mathrm{e}^{in\omega t}, \tag{10.48}$$

where the first term in (10.48) represents a backward precession, and the second a forward precession of the axis of the shaft. After substituting solution (10.48) into Eq. (10.46), we obtain for the amplitudes of direct and reverse precessions the equations

$$Nz_n - (n\omega)^2 \rho z_n - p_n = 0, \quad n = -1, 1. \tag{10.49}$$

The boundary conditions can be converted to the form

$$\begin{aligned} Q_n(0) &= (c_1 + in\omega\,\kappa_1)z_n(0) + (d_1 + in\omega\,\nu_1)\bar{z}_{-n}(0), \quad M_n(0) = 0, \\ Q_n(l) &= -(c_2 + in\omega\,\kappa_2)z_n(l) - (d_2 + in\omega\,\nu_2)\bar{z}_{-n}(l), \quad M_n(l) = 0. \end{aligned} \tag{10.50}$$

The coefficients $c_1, c_2, \kappa_1, \kappa_2, d_1, d_2, \nu_1, \nu_2$ for the left and right supports are determined in accordance with (10.45).

Thus, the amplitudes of the forward and backward precessions turn out to be connected by virtue of the boundary conditions. For their determination, we proceed as in the case of isotropic supports; namely, representing them in the form of

$$z_n(s) = \sum_{k=1}^{\infty} \beta_{nk}\tilde{z}_k(s) + \sum_{j=1}^{2} b_{nj}\hat{z}_j(s), \quad n = -1, 1. \tag{10.51}$$

where $b_{n1} = z_n(0), b_{n2} = z_n(l)$ are displacements of the ends of the shaft at the forward and backward precessions.

Using Eqs. (10.49), as in the case of isotropic supports, we get the algebraic equations for the coefficients β_{nk}, b_{nj}, from which we directly obtain

$$\beta_{nk} = \beta_{nk}(b_{nj}), \quad n = -1, 1, \quad j = 1, 2, \quad k = 1, 2.... \tag{10.52}$$

Performing the boundary conditions (10.50) for the forces $Q_n(0), Q_n(l)$ by integration, we obtain a system of four inhomogeneous algebraic equations of the form (10.38) for the unknowns $b_{11}, b_{12}, b_{-11}, b_{-12}$.

By solving these equations and then using (10.52) and (10.51), we can find the amplitudes of the direct and reverse precession and, further with the help of (10.48), the displacements in a stationary coordinate system. The form of solution (10.48) shows that motion at the center of the shaft is elliptical (i.e. generalized precession). Hence, the motion will be forward (i.e. in the direction of shaft rotation) if $|z_1(s)| > |z_{-1}(s)|$ and backward if $|z_1(s)| < |z_{-1}(s)|$.

Moving on to a rotating coordinate system, we have

$$w(s,t) = z(s,t)\mathrm{e}^{-i\omega t} = z_{-1}(s)\mathrm{e}^{-2i\omega t} + z_1(s). \tag{10.53}$$

It is evident from (10.53) that in a rotary coordinate system the shaft, along with its constant for the cross-section bending given, is subject to additional bending with a frequency double that of shaft rotation. In the case of isotropic supports the first term vanishes.

Therefore, the anisotropy of the supports brings about backward precession of the axis of the shaft. This is the essential part of the solution because, in the system of coordinates associated with the shaft, it changes over time and variable stresses in the shaft correspond to this part of the solution.

10.5 Parametric Oscillations of a Shaft with Twofold Bending Stiffness

The shafts of the rotors of some machines (in particular, turbogenerator rotors) have the peculiarity that their cross-section possesses different values of the principal moments of inertia. Consequently, the shafts are unequally stiff in the different planes of bending. If the shaft does not rotate, in each of its principal planes of bending there can occur independent oscillations with frequencies corresponding to the elastic properties of the shaft in the given plane. However, if the shaft rotates the oscillation frequencies are the frequencies of coupled vibrations of the shaft in both planes.

Let us consider the oscillations of a shaft with dual stiffness of bending under its own weight, which is essential, for example, for the heavy rotors of turbogenerators. The shaft will be assumed here as absolutely balanced and relying on absolutely rigid supports at its ends. Let J_u, J_v be the principal moments of inertia of the cross-section about the principal axes of inertia $u, v(J_u < J_v)$; and $\beta_u = \frac{1}{EJ_u}, \beta_v = \frac{1}{EJ_v}$ the bending compliances. We introduce the complex displacement $w(s)$ and bending moment $M_w(s)$ in the rotary coordinate system

$$w(s) = u(s) + iv(s), \;\; M_w(s) = M_v(s) + iM_u(s),$$

where $u(s), v(s)$ are displacements along the axes u, v; and $M_u(s), M_v(s)$ are bending moments about the same axes. Hence, in the stationary axes we have

$$z(s) = w(s)\mathrm{e}^{i\omega t}, \quad M(s) = M_w(s)\mathrm{e}^{i\omega t}. \tag{10.54}$$

The equations for bending in the principal (rotary) axes are of the form

$$-u'' = \beta_v M_v, \quad -v'' = \beta_u M_u. \tag{10.55}$$

We introduce the notation

$$\beta = \frac{\beta_u + \beta_v}{2}, \quad \Delta = \frac{\beta_u - \beta_v}{2} > 0,$$

where β is average compliance; and Δ is the different stiffness criterion. Then, Eqs. (10.55) can be rewritten as follows

$$\begin{aligned} -u'' - (\beta - \Delta)M_v = 0, \\ -v'' - (\beta + \Delta)M_u = 0. \end{aligned} \quad (10.56)$$

Multiplying the second Eq. (10.56) by the imaginary unit and adding to the first, we obtain the equation for bending in the rotary axes

$$-w'' - \beta M_w + \Delta \bar{M}_w = 0 \quad (10.57)$$

or in a stationary coordinate system

$$-z'' - \beta M + \Delta e^{2i\omega t} \bar{M} = 0. \quad (10.58)$$

Equation (10.58) contains, unlike the previous cases, a periodically varying coefficient; therefore, vibrations arising in the system will be parametric.

The complete system of equations describing vibrations of a twofold bending stiffness shaft on pivot supports under the weight load $q(s)$ is as follows

$$\begin{aligned} -M'' + \rho \ddot{z} - iq = 0, \\ -z'' - \beta M + \Delta e^{2i\omega t} \bar{M} = 0, \end{aligned} \quad (10.59)$$

$$M(0,t) = 0, \ M(l,t) = 0, \ z(0,t) = 0, \ z(l,t) = 0. \quad (10.60)$$

The solution to (10.59), (10.60) can be sought in the form

$$\begin{aligned} M(s,t) = i \sum_{n=0,2} M_n(s) e^{in\omega t}, \\ z(s,t) = i \sum_{n=0,2} z_n(s) e^{in\omega t}. \end{aligned} \quad (10.61)$$

After substituting (10.61) into both Eqs. (10.59) and conditions (10.60), and equating the coefficients of like powers of the exponents, we obtain two systems of coupled equations with respect to $M_0(s), M_2(s), z_0(s), z_2(s)$ with homogeneous conditions of the form (10.60)

$$\begin{aligned} -M_0'' - q = 0, \\ -z_0'' - \beta M_0 - \Delta \bar{M}_2 = 0, \end{aligned} \quad (10.62)$$

$$
\begin{aligned}
-M_2'' - (2\omega)^2 \rho z_2 &= 0, \\
-z_2'' - \beta M_2 - \Delta \bar{M}_0 &= 0.
\end{aligned} \tag{10.63}
$$

The solution to Eqs. (10.62), (10.63) can be found in the following sequence:

(a) from the first Eq. (10.62) with zero conditions for the bending moment at the shaft ends, we find $M_0(s)$;
(b) at known function $M_0(s)$ and zero boundary conditions for $M_2(s), z_2(s)$, we construct the solution to Eqs. (10.63);
(c) now that we know $M_0(s)$ and $M_2(s)$, we find $z_0(s)$ from the second Eq. (10.62);
(d) using formulas (10.61), we obtain the complete solution to the problem.

So, particularly, we have

$$
z(s,t) = iz_0(s) + iz_2(s)e^{i2\omega t}. \tag{10.64}
$$

10.6 Vibrations of the Blade of a Steam Turbine Subject to Centrifugal Forces

The problem of free oscillations of the blade of a steam turbine subject to the centrifugal forces that arise at rotation of the turbine rotor. We are talking about vibrations of the longest and most flexible blades of the last stages of large turbines. The action of centrifugal forces can significantly impact, in particular, the forms and frequencies of the free oscillations of these blades.

Let us consider the blade as a thin naturally twisted rod of variable cross-section and ignore the initial static curvature of the axis of the rod (i.e. assume the blade is a straight rod). It is specified that the blade is fixed at one end in an absolutely hard disk while the other end of the blade is free (for a more general formulation of the problem of vibrations of naturally twisted blades connected among each other by a bandage, see [1]). Along the axis of the blade direct the radial coordinate r, which varies for the blade in the range from R to $R + l$, where R is the radius of the disk; and l is length of the blade. In an arbitrary cross-section of the blade we introduce an orthogonal coordinate system 1, 2. Let us assume that the disk has a plane of symmetry and that axis 1 lies in this plane, while axis 2 is perpendicular to the plane $1, r(1, 2, r$ is a right-hand coordinate system). The notion of a naturally twisted rod means that the principal axes of inertia of the rod cross-section slowly change their angular position in the plane 1, 2 when changing the coordinate r. Natural twisting of the blade is carried out during manufacture.

The centrifugal force of an infinitely thin layer of a blade perpendicular to the axis r is equal to

$$
dC = \omega^2 \rho s ds, \tag{10.65}
$$

where s is the same coordinate as r; and ω is the rotational frequency of the rotor.

Bending the blade in the plane $2, r$ results in an elementary centrifugal force $dC(s)$, which is applied to the blade cross-section with coordinate s, and an elementary bending moment in the cross-section of coordinate r equal to $dC(s)(u_2(s) - u_2(r))$. In turn, bending the blade in the plane $1, r$ results in just the same elementary force $dC(s)$ and the appearance of elementary moment $dC(s)\left(u_1^*(s, r) - u_1(r)\right), u_1^*(s, r) = u_1(s)\frac{r}{s}$. Multiplier $\frac{r}{s}$ appears in this formula because the force $dC(s)$, directed along the radius, is not parallel to the axis of the blade r; that is, the radius passing through the center of the cross-section of the blade with the coordinate s and the axis of the blade r form an albeit small angle with the vertex at the center of rotation of a disk with blades.

Therefore, the arm of the force $dC(s)$ in cross-section with coordinate r is $\frac{r}{s}$ times less than the same arm in the cross-section of coordinate s. The complete bending moments caused by centrifugal forces of the blade portion with coordinate $s, r \le s \le R + l$, will be equal to

$$
\begin{aligned}
M_1^c &= \int\limits_r^{R+l} dC(s)\left(u_1^*(s, r) - u_1(r)\right), \quad u_1^*(s, r) = u_1(s)\frac{r}{s}, \\
M_2^c &= \int\limits_r^{R+l} dC(s)(u_2(s) - u_2(r)).
\end{aligned}
\tag{10.66}
$$

Corresponding distributed moments can be determined by the formulas

$$
\begin{aligned}
m_1^c &= \frac{dM_1^c}{dr} = \frac{d}{dr}\int\limits_r^{R+l} dC(s)\left(u_1(s)\frac{r}{s} - u_1(r)\right) \\
&= \omega^2 \int\limits_r^{R+l} \rho(s)u_1(s)ds + C\vartheta_1, \\
m_2^c &= \frac{dM_2^c}{dr} = \frac{d}{dr}\int\limits_r^{R+l} dC(s)(u_2(s) - u_2(r)) = -C\frac{du_2}{dr} = C\vartheta_2,
\end{aligned}
\tag{10.67}
$$

where both formula (10.65) and formulas $\vartheta_1 = -\frac{du_1}{dr}, \vartheta_2 = -\frac{du_2}{dr}$ have been used.

Index 1 is assigned to all the quantities that refer to deformations of the blade in the plane $1, r$, and index 2 is ascribed to those that relate to the plane $2, r$.

Entering $u = (u_1, u_2), \vartheta = (\vartheta_1, \vartheta_2)$, $Q = (Q_1, Q_2), M = (M_1, M_2), m = (m_1^c, m_2^c)$ and the matrix of elasticity $\beta = (\beta_{i,j}), i,j = 1, 2$, analogously with Eqs. (1.48), we can write a system of equations describing the free vibrations of a naturally twisted blade of a steam turbine subject to centrifugal force

$$\begin{aligned} -\frac{\mathrm{d}Q}{\mathrm{d}x} - \lambda^2 \rho u = 0, &\quad \frac{\mathrm{d}u}{\mathrm{d}x} + \vartheta = 0, \\ Q - \frac{\mathrm{d}M}{\mathrm{d}x} - m = 0, &\quad \frac{\mathrm{d}\vartheta}{\mathrm{d}x} - \beta M = 0. \end{aligned} \tag{10.68}$$

An approximate solution of the system of Eqs. (10.68) can be efficiently obtained with the help of the projection method, which is based on the mixed variational principle (Chap. 4). As was the case with the coordinate functions, it is convenient to choose the forms of the free oscillations of a rod of constant cross-section, clamped at one end and free at the other (Chap. 7). More detail on the problem of the oscillations of a blade subject to centrifugal force is beyond the scope of this chapter.

10.7 Joint Vibrations of the Disk and Blades of a Turbine Rotor

Let us consider the joint free vibrations of a disk that does not rotate and blades. We examine the disk as a round plate of variable thickness clamped in its center, and the blades as rods of variable cross-section rigidly connected to the disk in their root sections and free at their other ends. The natural twisting of blades is not taken into account here.

Equations for the free bending vibrations of a rod of variable cross-section were considered in Chap. 1 and can be written as

$$\begin{aligned} D^{(1)}\xi^{(1)} - \lambda^2 R^{(1)}\eta^{(1)} &= 0, \\ D^{(1)*}\eta^{(1)} - B^{(1)}\xi^{(1)} &= 0, \end{aligned} \tag{10.69}$$

where

$$\xi^{(1)} = (Q, M), \quad \eta^{(1)} = (u, \vartheta), \tag{10.70}$$

Q, M is the shearing force and bending moment; and u, ϑ is the displacement of the axis of the rod and rotation angle of the cross-section. Such operators and their properties were considered in Chap. 1.

For simplicity, we consider the case when the vibrations of the disk occur in the axial direction and are axisymmetric, while the blades oscillate synchronously with the disk. Equations for the axisymmetric free oscillations of a circular plate can also be written in operator form (Chap. 7)

$$\begin{aligned} D^{(2)}\xi^{(2)} - \lambda^2 R^{(2)}\eta^{(2)} &= 0, \\ D^{(2)*}\eta^{(2)} - B^{(2)}\xi^{(2)} &= 0. \end{aligned} \tag{10.71}$$

The generalized forces and displacements would then be

$$\xi^{(2)} = (Q_r, M_r, M_\varphi), \quad \eta^{(2)} = (w, \varphi), \tag{10.72}$$

where Q_r is the shearing force per unit circumference of radius $r; M_r, M_\varphi$ are the bending moments per unit length with M_r applying to the tangential sections and M_φ to diametrical sections; w is deflection of the plate; and φ is the slope angle of the tangent plane.

An approximate solution to the problem of finding the frequencies and forms of free joint oscillations of a disk and blades can be obtained as follows:

(1) We can solve the problem of the free oscillations of a separate blade under simple homogeneous boundary conditions (i.e. when one end is clamped and the other is free)

$$\tilde{\eta}^{(1)}(0) = 0, \quad \tilde{\xi}^{(1)}(l) = 0, \tag{10.73}$$

where l is length of the blade. These solutions can be represented as

$$\tilde{\eta}^{(1)}(r) = \sum_{k=1}^{n} \beta_k^{(1)} \tilde{\eta}_k^{(1)}(r), \quad \tilde{\xi}^{(1)}(r) = \sum_{k=1}^{n} \alpha_k^{(1)} \tilde{\xi}_k^{(1)}(r), \tag{10.74}$$

where r is a coordinate counted along the axis of the blade and directed along the radius of the disk; and $\tilde{\eta}_k^{(1)}, \tilde{\xi}_k^{(1)}$ are the forms of free oscillations of the blade much like the rod of a variable cross-section under conditions of the form (10.73). They can be determined using the iterative method.

Then, a solution for the blade as part of the disk-blade system can be represented as follows

$$\eta^{(1)}(r) = \tilde{\eta}^{(1)}(r) + \widehat{\eta}^{(1)}(r), \quad \xi^{(1)}(r) = \tilde{\xi}^{(1)}(r), \tag{10.75}$$

$$\widehat{\eta}^{(1)}(r) = b_1 \widehat{\eta}_1^{(1)}(r) + b_2 \widehat{\eta}_2^{(1)}(r), \tag{10.76}$$

$$\widehat{\eta}_1^{(1)} = \begin{pmatrix} 1 - \frac{r}{l} \\ \frac{1}{l} \end{pmatrix}, \quad \widehat{\eta}_2^{(1)} = \begin{pmatrix} \frac{r}{l} \\ -\frac{1}{l} \end{pmatrix}, \tag{10.77}$$

where the coefficients $\alpha_k^{(1)}, \beta_k^{(1)}, k = 1, \ldots, n, \quad b_1, b_2$ remain unknown.

(2) We can solve the problem of the vibrations of a disk under simple boundary conditions; namely

$$\tilde{\eta}^{(2)}(0) = 0, \quad \tilde{X}^{(2)}(R) = 0, \tag{10.78}$$

where $X^{(2)}(R) = (2\pi R Q_r(R), 2\pi R M_r(R))$.

These solutions can be presented as expansions

$$\tilde{\eta}^{(2)}(r) = \sum_{k=1}^{m} \beta_k^{(2)} \tilde{\eta}_k^{(2)}(r), \quad \tilde{\xi}^{(2)}(r) = \sum_{k=1}^{m} \alpha_k^{(2)} \tilde{\xi}_k^{(2)}(r), \tag{10.79}$$

where we choose the forms of free oscillations of a disk under conditions of type (10.78) for $\tilde{\eta}_k^{(2)}, \tilde{\xi}_k^{(2)}$.

For a disk that is part of the disk-blade system, we can write the solution in the form

$$\eta^{(2)}(r) = \tilde{\eta}^{(2)}(r), \quad \xi^{(2)}(r) = \tilde{\xi}^{(2)}(r) + \widehat{\xi}^{(2)}(r), \tag{10.80}$$

$$\widehat{\xi}^{(2)}(r) = a_1 \widehat{\xi}_1^{(2)}(r) + a_2 \widehat{\xi}_2^{(2)}(r), \tag{10.81}$$

where a_1, a_2 are the unknown coefficients. For $\widehat{\xi}_1^{(2)}, \widehat{\xi}_2^{(2)}$ we take the solution of the relevant static problems:

(a) for $\widehat{\xi}_1^{(2)}(r)$ under the conditions

$$\widehat{w}_1(0) = 0, \widehat{\varphi}_1(0) = 0, \widehat{Q}_{r1}(R) = 1, \quad \widehat{M}_{r1}(R) = 0, \tag{10.82}$$

(b) for $\widehat{\xi}_2^{(2)}(r)$ under the conditions

$$\widehat{w}_2(0) = 0, \quad \widehat{\varphi}_2(0) = 0, \quad \widehat{Q}_{r2}(R) = 0, \quad \widehat{M}_{r2}(R) = 1. \tag{10.83}$$

(3) We can solve the problem of the joint vibrations of disk and blades described by the equations

$$\begin{aligned} D\xi - \lambda^2 R\eta &= 0, \\ D^*\eta - B\xi &= 0, \end{aligned} \tag{10.84}$$

where λ is the frequency of joint oscillations; and ξ, η are the forms of free oscillations of the system as a whole.

Further, we apply the projection method to all these equations. This means multiplying them scalarly by $\tilde{\eta}_k^{(1)}, \tilde{\xi}_k^{(1)}$ (for the blade) and by $\tilde{\eta}_k^{(2)}, \tilde{\xi}_k^{(2)}$ (for the disk). After substituting the expressions (10.74)–(10.77), (10.79)–(10.81) for $\eta^{(1)}, \xi^{(1)}, \eta^{(2)}, \xi^{(2)}$ into these projection conditions, and using the properties of orthonormality of the individual forms of the free oscillations of the blades and the disk we get a system of $2(n+m)$ algebraic equations, from which such relations are found

$$\alpha_k^{(1)} = \alpha_k^{(1)}(b_1, b_2), \quad \beta_k^{(1)} = \beta_k^{(1)}(b_1, b_2), \quad k = 1, 2, \ldots, n,$$
$$\alpha_k^{(2)} = \alpha_k^{(2)}(a_1, a_2), \quad \beta_k^{(2)} = \beta_k^{(2)}(a_1, a_2), \quad k = 1, 2, \ldots, m.$$

Therefore, by virtue of formulas (10.74)–(10.77), (10.79)–(10.81), the forces and displacements in any cross-section of both blade and disk can be expressed by the four unknowns a_1, a_2, b_1, b_2. Since the main boundary conditions at the ends of the blades and at the disk center have already been implemented, we use four conditions of the joint motion of both the disk and the blades where they connect for these unknowns

$$\begin{aligned} pQ(0) = 2\pi RQ_r(R), \quad pM(0) = 2\pi RM_r(R), \\ u(0) = w(R), \quad \vartheta(0) = \varphi(R), \end{aligned} \tag{10.85}$$

where p is the number of blades connected to the disk.

Equations (10.85), in which the forces and displacements are expressed by a_1, a_2, b_1, b_2, represent a system of four homogeneous algebraic equations of the form

$$Ax = 0, \tag{10.86}$$

where $x = (a_1, a_2, b_1, b_2)$; and the elements of matrix A depend on the frequency $a_{i,j} = a_{i,j}(\lambda^2), i,j = 1, .., 4$. Let us not forget that we cannot distinguish between row vectors and column vectors in the record. The solvability condition of Eqs. (10.86) is perfect for determining the frequencies λ_k of free oscillations, while the solutions of these equations enable us to construct the forms of joint oscillations of the disk and blades.

The technique we have used to solve the problem of the joint vibrations of the disk and blades of a steam turbine (in the simplified formulation) is called the "decomposition method". When applied to the problems of dynamics of machines and structures, this method has the advantage of being divisible into simple elements, for each of which it is possible to build a relatively simple solution at arbitrary values of the unknown quantities on the boundaries of these elements (decomposition) when the vibrations of a complex mechanical system are considered. Further, these values on the boundaries can be determined from the conjugation conditions of all the elements, which allows us to construct a solution to the problem as a whole (composition). Details on application of the decomposition method can be found in [2], where the problem of the vibrations of a multisupport rotating shaft (such as the rotor shaft of a turbine set unit for a thermal or nuclear power plant) is considered. A solution to the problem of the vibrations of spatial construction of a turboset can be found in [3].

References

1. Fridman, V. M. (1965). Bending vibrations of a package of naturally twisted rods subject to centrifugal forces. In *Proceedings of the Leningrad Polytechnic Institute. Dynamics and strength of machines*, *252*, 87–94 (in Russian).
2. Soussie, I. R., & Fridman, V. M. (1977). Vibrations and balancing a shaft with distributed parameters on anisotropic elastic-damping bearings. Academy of sciences of the U.S.S.R. machine. *Science, 6,* 24–31. (in Russian).
3. Fridman, V. M. (1980). Turbounit foundation vibrations. *Mechanical engineering, vibrations in technique handbook* (Vol. 3, pp. 532–537). Moscow (in Russian).

Chapter 11
Stability of the Equilibrium Position of a Rotating Shaft Axis

The present chapter is devoted to methods used to investigate the stability of the static position of an elastic body by giving examples of particular tasks. We consider the stability of the axis of an absolutely balanced shaft rotating at frequency ω with a round cross-section on anisotropic elastic-damping supports. Equations for the bending vibrations of the shaft and the boundary conditions can be written as

$$Nz + R\ddot{z} = 0, \tag{11.1}$$

$$\begin{aligned} Q(0,t) &= P_1(z,\dot{z}), \quad M(0,t) = 0, \\ Q(l,t) &= -P_2(z,\dot{z}), \quad M(l,t) = 0, \end{aligned} \tag{11.2}$$

where $P_1(z,\dot{z})$ and $P_2(z,\dot{z})$ are oil film reactions.

Let us consider small deviations $\delta z(s,t)$ of the shaft from the position of its static equilibrium such that

$$z(s,t) = z_0(s) + \delta z(s,t), \tag{11.3}$$

where $z_0(s)$ is a solution of the static problem

$$Nz_0 = 0, \tag{11.4}$$

$$\begin{aligned} Q_0(0) &= P_1(z_0,0), \quad M_0(0) = 0, \\ Q_0(l) &= -P_2(z_0,0), \quad M_0(l) = 0. \end{aligned} \tag{11.5}$$

After substituting solution (11.3) into (11.1) and expanding the functions $P_1(z,\dot{z})$ and $P_2(z,\dot{z})$ in the Taylor series in the neighborhood of the static equilibrium position $z = z_0, \dot{z} = 0$ but retaining the linear summands, then taking (11.4) and (11.5) into account the equation and boundary conditions subject to variations take the form

V. Fridman, *Theory of Elastic Oscillations*, Foundations of Engineering Mechanics, DOI 10.1007/978-981-10-4786-2_11

$$N\delta z + R\delta\ddot{z} = 0, \tag{11.6}$$

$$\begin{aligned} \delta Q(0,t) &= c_1\delta z(0,t) + d_1\delta\bar{z}(0,t) + \kappa_1\delta\dot{z}(0,t) + \nu_1\delta\dot{\bar{z}}(0,t), \quad \delta M(0,t) = 0, \\ \delta Q(l,t) &= -(c_2\delta z(l,t) + d_2\delta\bar{z}(l,t) + \kappa_2\delta\dot{z}(l,t) + \nu_2\delta\dot{\bar{z}}(l,t)), \quad \delta M(l,t) = 0. \end{aligned} \tag{11.7}$$

In Chap. 10 we studied the forced oscillations of a round shaft on anisotropic supports. Setting the unbalanced centrifugal forces equal to zero, we arrive at a corresponding homogeneous problem for which Eq. (11.1) remains valid and such boundary conditions as

$$\begin{aligned} Q(0,t) &= c_1 z(0,t) + d_1\bar{z}(0,t) + \kappa_1\dot{z}(0,t) + \nu_1\dot{\bar{z}}(0,t), \quad M(0,t) = 0, \\ Q(l,t) &= -(c_2 z(l,t) + d_2\bar{z}(l,t) + \kappa_2\dot{z}(l,t) + \nu_2\dot{\bar{z}}(l,t)), \quad M(l,t) = 0. \end{aligned} \tag{11.8}$$

can be added.

Comparing the variational Eqs. (11.6), (11.7) with the equations for oscillations of a round balanced shaft (11.1), (11.8) on linear anisotropic supports, their full natural identity becomes clear. This means that the way in which the solution to problems (11.1), (11.8) behaves, it is possible to judge the stability of the static position of the axis of a balanced rotary shaft.

Let us write the solution to Eq. (11.1) in the form

$$z(s,t) = z_1(s)e^{ht} + z_{-1}(s)e^{\bar{h}t}, \tag{11.9}$$

where $h = q + i\lambda$. Studying the stability of the equilibrium position of a rotating shaft axis comes down to scrutiny of the signs of q and real parts of the unknown complex numbers h.

Just as in the problem of forced vibrations of a shaft on anisotropic supports, the functions $z_n(s), n = -1, 1$, can be sought in the form

$$z_n(s) = \sum_{k=1}^{\infty} \beta_{nk}\tilde{z}_k(s) + \sum_{j=1}^{2} b_{nj}\hat{z}_j(s), \quad n = -1, 1. \tag{11.10}$$

By virtue of Eq. (11.1), the coefficients $\beta_{nk}, n = -1, 1, k = 1, 2, \ldots$, can be expressed (just as before) by the coefficients $b_{nj}, n = -1, 1, j = 1, 2$, which after the boundary conditions (11.8) are carried out can be determined from the system of four homogeneous algebraic equations

$$Ax = 0, \tag{11.11}$$

where $x = (b_{-11}, b_{-12}, b_{11}, b_{12})$ is the unknown vector; and $A = A(q, \lambda)$ is the matrix with complex coefficients. The condition of existence of a nonzero solution to (11.11) can be written as the vanishing of determinant Δ of the matrix A

$$\Delta_1(q,\lambda) = 0, \quad \Delta_2(q,\lambda) = 0, \quad \Delta = \Delta_1 + i\Delta_2. \tag{11.12}$$

Evaluation of the values q found from (11.12) allows us to judge the stability of the equilibrium position of the axis of a rotating shaft.

11.1 Methods Used to Investigate Stability

1. The direct solution to Eq. (11.12).

We are confined in expansions (11.10) by a finite number of terms $k = 1, 2, \ldots, m$. Then, solving nonlinear equations (11.12) directly makes it necessary to receive all the roots $q_1, \lambda_1, q_2, \lambda_2, \ldots, q_m, \lambda_m$, and then evaluate the signs of all q. Should $q_k > 0$ for at least one root of Eq. (11.12), then the static position of the shaft is unstable.

2. The frequency criterion for the study of stability (Mikhailov criterion).

Let us suppose that $q = 0$ such that $h = i\lambda$ (i.e. the system carries out purely harmonic oscillations at the boundary of the stability region). We now construct the dependences of $\Delta_1(0,\lambda), \Delta_2(0,\lambda)$ on λ and compare them. If the roots of functions $\Delta_1(0,\lambda)$ and $\Delta_2(0,\lambda)$ strictly alternate, then the equilibrium position is stable. If they do not, then this means that $q_k > 0$ for at least one root of Eq. (11.12) and the system is unstable. Here $\Delta_1(0,\lambda)$ and $\Delta_2(0,\lambda)$ can be built by repeatedly using an algorithm to solve the problem of the free oscillations of a shaft on anisotropic supports for each λ, and hence there is no requirement to solve the system of Eq. (11.12).

3. The parametric method (determining the boundary of stability).

An oscillatory system can have parameters whose value determines whether the system is stable or unstable. Such a parameter in the problem of vibrations of a shaft on sliding bearings can for example be temperature (i.e. dependent on the viscosity of oil fed into the bearing). Shaft rotation frequency is the most important parameter determining the stability of the static position of a shaft axis. In reality, it has been discovered that the stability or instability of the equilibrium position of a rotating shaft on anisotropic supports can significantly depend on the frequency of its rotation. At a certain value of frequency $\omega > \omega^*$ there occurs a loss of stability of the static position of the axis of the shaft, which entails the emergence of self-oscillations (auto-oscillations), even if the shaft is balanced. The value ω^* is called the "boundary of stability" of the given parameter.

Let us try to determine the boundary of the stability domain in the problem of vibrations of a shaft on anisotropic sliding bearings. We proceed from the fact that the elastic-damping characteristics of oil-lubricated bearings significantly depend on the speed of rotation of the shaft ω. We regard the shaft rotational frequency ω as the parameter on which these characteristics essentially depend and, as a consequence, so does the stability of the static shaft position. Let us assume we are located at the regional boundary of stability when $q = 0$. Hence, the conditions of existence of a nonzero solution of the system (11.11) will take the form

$$\Delta_1(\omega, 0, \lambda) = 0, \quad \Delta_2(\omega, 0, \lambda) = 0. \tag{11.13}$$

Let us further assume that we are interested in the range $\omega_1 \leq \omega \leq \omega_2$ and that the position of the rotating shaft is stable at $\omega = \omega_1$. Looking at the parameter values ω from the indicated range, we seek a value of this parameter ω^* at which both Eqs. (11.13) are fulfilled. The minimal value $\omega^*_{\min}$ at which this condition is satisfied is called the "boundary of stability", and the corresponding value $\lambda = \lambda(\omega^*_{\min})$ represents the frequency of auto-oscillations at the stability boundary. More detail on this theme can be found in [1], where the problem of the stability boundary of the axis position of a rotating shaft simply supported at both ends on the bearings with liquid lubricant has been solved. Calculations of the stability boundaries for the rotor shaft of a high-power turbogenerator have been carried out. It has been assumed that the shaft is symmetric relative to the middle of its length and that both bearings are identical. We succeed here in excluding the parameter λ from the system of Eq. (11.13) and to obtain a single equation

$$f(\omega) = 0 \tag{11.14}$$

to determine the boundaries of sustainability $\omega = \omega^*$. The dependences $f_i(\omega), i = 1, 2, 3$ have been calculated at different values of the radial clearance $\delta_1 < \delta_2 < \delta_3$. Figure 11.1 depicts the function $f(\omega)$ calculated at the lowest value of the gap equal to δ_1.

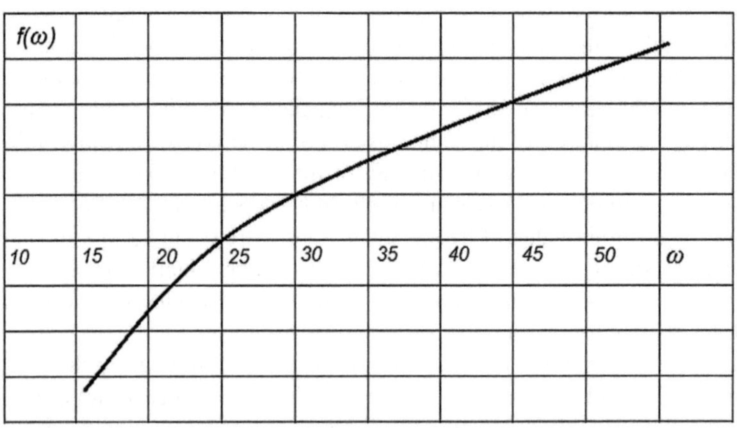

Fig. 11.1 Determination of the stability boundary of a rotating shaft (dimensions along the abscissa axis are given in hertz)

The corresponding values of the roots of Eq. (11.14) are given in hertz such as $\omega_1^* = 25; \omega_2^* = 46.7; \omega_3^* = 59.1$. This means that, as a result of calculations throughout the entire frequency range of rotor rotation from $\omega = 0$ to the operating rotational speed $\omega = 50$ at the largest radial clearance equal to δ_3, the position of the axis of the rotating shaft of the rotor is stable.

4. The iterative stability criterion [2].

We can write the equation for the bending vibrations of a balanced round shaft on anisotropic elastic-damping supports in the operator form and introduce boundary conditions into the equation

$$Nz + R\ddot{z} + G\dot{z} + Cz = 0, \quad (11.15)$$

where

$$z = \begin{pmatrix} x \\ y \end{pmatrix}, \quad R = \begin{pmatrix} \rho & 0 \\ 0 & \rho \end{pmatrix}, \quad N = \begin{pmatrix} \frac{\partial^2}{\partial s^2} EJ \frac{\partial^2}{\partial s^2} & 0 \\ 0 & \frac{\partial^2}{\partial s^2} EJ \frac{\partial^2}{\partial s^2} \end{pmatrix}. \quad (11.16)$$

We further introduce the operators C and G that characterize the elastic-damping properties of the supports

$$\begin{aligned} C &= \begin{pmatrix} c_{xx} & c_{xy} \\ c_{yx} & c_{yy} \end{pmatrix}|_{s=0}\delta(s-0) + \begin{pmatrix} c_{xx} & c_{xy} \\ c_{yx} & c_{yy} \end{pmatrix}|_{s=l}\delta(s-l), \\ G &= \begin{pmatrix} \kappa_{xx} & \kappa_{xy} \\ \kappa_{yx} & \kappa_{yy} \end{pmatrix}|_{s=0}\delta(s-0) + \begin{pmatrix} \kappa_{xx} & \kappa_{xy} \\ \kappa_{yx} & \kappa_{yy} \end{pmatrix}|_{s=l}\delta(s-l). \end{aligned} \quad (11.17)$$

We denote $p = R\dot{z}$ and rewrite Eq. (11.15) as the system of equations

$$\begin{aligned} R^{-1}p - \dot{z} &= 0, \\ -(N+C)z - GR^{-1}p - \dot{p} &= 0. \end{aligned} \quad (11.18)$$

By introducing the new unknown φ and operator A

$$\varphi = \begin{pmatrix} z \\ p \end{pmatrix}, \quad A = \begin{pmatrix} 0 & R^{-1} \\ -(N+C) & -GR^{-1} \end{pmatrix},$$

We can rewrite the system (11.18) in such a way

$$A\varphi - \dot{\varphi} = 0. \quad (11.19)$$

The solution to Eq. (11.19) can be represented in the form of an exponential function with an as-yet unknown coefficient of degree h

$$\varphi(s,t) = \varphi(s)e^{ht}. \tag{11.20}$$

By virtue of formulas (11.19), (11.20) there will be

$$A\varphi - h\varphi = 0, \quad \varphi = \varphi(s), \tag{11.21}$$

where $h = h_k = q_k + i\lambda_k, k = 1, 2, \ldots$ are eigenvalues of the operator A.

When there is asymptotic stability of the equilibrium position for all $k = 1, 2, \ldots$ the inequalities $q_k < 0$ need to be carried out.

We transform Eq. (11.21) by first noting that:

(a) if h is an eigenvalue of the operator A, then $(h - \omega)$ is an eigenvalue of the operator $(A - \omega E)$ and $(h + \omega)$ an eigenvalue of the operator $(A + \omega E)$

$$(A - \omega E)\varphi - (h - \omega)\varphi = 0, \tag{11.22}$$

$$(A + \omega E)\varphi - (h + \omega)\varphi = 0, \tag{11.23}$$

(b) the inverse operator has eigenvalues that are inverse to the eigenvalues of the direct operator

$$(A - \omega E)^{-1}\varphi - (h - \omega)^{-1}\varphi = 0. \tag{11.24}$$

From (11.23) and (11.24) it follows that

$$(A - \omega E)^{-1}(A + \omega E)\varphi - (h - \omega)^{-1}(h + \omega)\varphi = 0. \tag{11.25}$$

So, instead of (11.21) we can write the equation as

$$B\varphi - \nu\varphi = 0, \quad \varphi = \varphi(s), \tag{11.26}$$

where

$$\begin{aligned} B &= (A - \omega E)^{-1}(A + \omega E); \\ \nu &= \frac{h + \omega}{h - \omega}; \quad h = q + i\lambda; \end{aligned} \tag{11.27}$$

ν is an eigenvalue of the operator B; and ω is any positive number.

According to formula (11.27)

$$|\nu|^2 = \frac{(q+\omega)^2 + \lambda^2}{(q-\omega)^2 + \lambda^2}, \quad \omega > 0. \tag{11.28}$$

Analyzing formula (11.28), it is not hard to see that

at $q<0$ $|\nu|<1$ is asymptotic stability;

at $q=0$ $|\nu|=1$ is the boundary of the domain of stability;

at $q>0$ $|\nu|>1$ is instability.

Thus, if all eigenvalues h_k of the operator A lie in the left half-plane, then all eigenvalues ν_k of the operator B lie inside the unit circle. That is, by means of formula (11.28), the left half-plane for the parameter q is displayed on the inside of the unit circle for the parameter ν.

The eigenvalue with the largest modulus $r = |\nu|_{\max}$ of the operator B is called its "spectral radius". Hence, to judge asymptotic stability, it is important to establish that the spectral radius

$$r = |\nu|_{\max} < 1. \tag{11.29}$$

Direct verification of condition (11.29) in the problem of vibrations of a shaft with distributed parameters on anisotropic supports appears to be difficult. We therefore propose using an iterative process

$$\varphi_n = B\varphi_{n-1} \tag{11.30}$$

or

$$(A - \omega E)\varphi_n = f_n, \quad f_n = (A + \omega E)\varphi_{n-1}, \quad \omega > 0. \tag{11.31}$$

which is the same.

In other words, the stability of the static position of the elastic line of a shaft can be reduced to multiple solution of the problem of forced periodic oscillations of a shaft with positive rotational speed ω.

The stability of a static shaft position can be estimated by descending or ascending the norms of the successive approximations $\|\varphi_n\|$ at $n \to \infty$:

(1) If

$$\|\varphi_n\| \to 0 \tag{11.32}$$

independently of the choice of φ_0, then we have asymptotic stability of the equilibrium position.

(2) If there exists a φ_0 at which

$$\|\varphi_n\| \to \infty, \tag{11.33}$$

then the equilibrium position is unstable.

(3) At

$$\|\varphi_n\| \to \text{const} \tag{11.34}$$

we have located the regional boundary of stability.

In actual fact, let

$$\|\varphi_n\| = \|B\varphi_{n-1}\| = \|B^n\varphi_0\| < \varepsilon \tag{11.35}$$

at any preassigned small number ε and $n(\varepsilon)$. Since φ_0 is any element from the domain of definition of the operator B, then (11.35) is only feasible if

$$\|B^n\| \leq \varepsilon. \tag{11.36}$$

According to the theorem of Gelfand [3]

$$\|B^n\| > r^n, \tag{11.37}$$

where r is the spectral radius of the operator B. According to (11.36), (11.37) $r^n \to 0$, which means that $r < 1$ (i.e. the equilibrium position is asymptotically stable).

By contrast, let $r > 1$. Choose small $\varepsilon < 1$ and $n(\varepsilon)$ such that $r^n > \frac{1}{\varepsilon}$. Then, by virtue of (11.37), $\|B^n\| > \frac{1}{\varepsilon}$ and φ_0 can be specified such that there will be

$$\|\varphi_n\| = \|B^n\varphi_0\| \geq \frac{1}{\varepsilon}, \tag{11.38}$$

(i.e. $\|\varphi_n\| \to \infty$).

When neither (11.32) nor (11.33) are satisfied the situation corresponds to the boundary case, where there is stability but it is not asymptotic.

The advantage of the method outlined lies in the fact that it allows stability in problems with many degrees of freedom to be investigated, especially when the problem of direct determination of all the roots of Eq. (11.12) appears to be very laborious.

There is an extensive literature devoted to consideration of various stability criteria; hence, this subject will not be discussed further. The more detailed presentation of the iterative stability criterion given here is explained, in particular, by the fact that this book is mainly devoted to variational and iterative methods.

References

1. Soussie, I. R., & Fridman, V. M. (1968). Research into the stability of variable section shaft motion on sliding bearings using the modified Galerkin method. *Academy of Sciences of the U.S.S.R. Machine Science, 4*, 21–27 (in Russian).
2. Fridman, V. M. (1994). Nonlinear problems of power machine dynamics. In *International Symposium on the Active Control of Vibration, London*, pp. 179–187.
3. Krasnosel'skii, M. A., Vainikko, G. M., & Zabreiko, P. P. et al. (1972). *Approximate solution of operator equations* (496 pp). Berlin: Springer.

Chapter 12
Vibrations of an Internal Combustion Engine

12.1 Statement of the Problem (Original Model of an Internal Combustion Engine)

The problem of forced nonlinear periodic vibrations of an internal combustion engine (ICE) is considered [1]. The widespread design of the piston ICE involves a system of three shafts interconnected by gear wheels. These shafts are the crankshaft and two camshafts. In stationary operation mode the ICE shafts and gear wheels perform periodic torsional vibrations. The vibrations arise under the action of periodic torques that are generated by crank mechanisms. In turn, these mechanisms are driven by reciprocating motion of the pistons inside the ICE cylinders.

Let us begin with that part of an ICE consisting of shafts and gears, where the acting torques are assumed to be those given. The initial structure with distributed parameters can be substituted by the finite-dimensional system. For this purpose each of the shafts can be represented as a system of absolutely hard disks interconnected by torsion bars subject to torsion. The gear wheels are also assumed as disks, the resilient connection between which is determined by the elasticity of the teeth. The damper bodies and the object (i.e. the reactive load) driven by a motor are also considered additional disks. The total mechanical system can be conventionally shown as in Fig. 12.1. For descriptive reasons the image is given in planform. The hard disks are system nodes, while the torsion bars are replaced by weightless elastic springs. A scheme of the gearwheels is shown in Fig. 12.2.

There are two reasons for the difficulty of calculation of ICE oscillations. First, the oscillatory system under consideration has a complicated branched structure. Second, and very importantly, there are backlashes in the connections between the gears without which it is impossible to collect the system of gearwheels. In other words, the connections between the gears are discontinuous. The presence of backlashes in the tooth connections restricts the problem of ICE vibrations to being a nonlinear one.

V. Fridman, *Theory of Elastic Oscillations*, Foundations of Engineering Mechanics, DOI 10.1007/978-981-10-4786-2_12

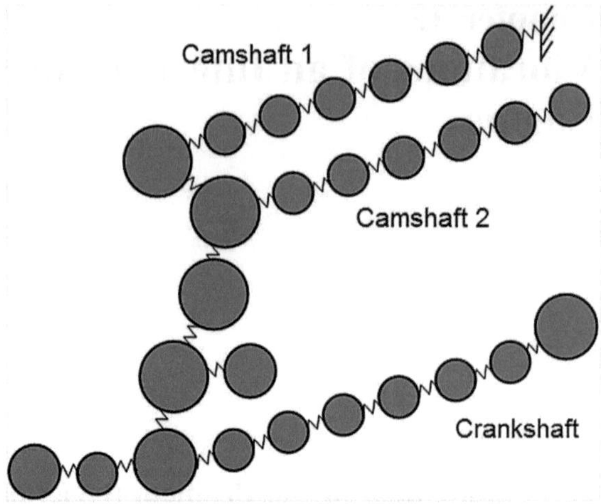

Fig. 12.1 Dynamic model of an internal combustion engine (ICE)

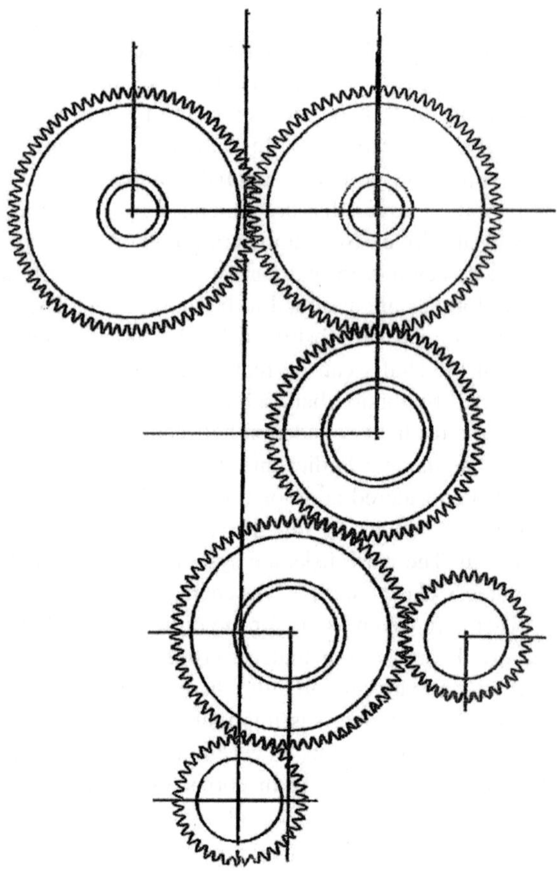

Fig. 12.2 Scheme of an ICE gearbox

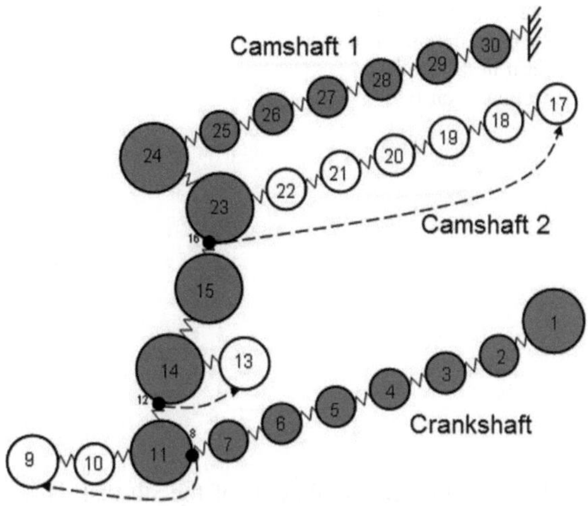

Fig. 12.3 Dynamic model of an ICE (calculation scheme)

12.2 The Method Used for Solution

Initially, the main backbone chain is set apart from the complex ramified elastic-mass system. It is indicated in Fig. 12.3 by black circles, whereas the lateral branches of the basic chain are indicated by white circles.

I

The nodes to which the branches are attached are set apart in the main chain. The "fictitious" nodes with very small mass are attached to these nodes. Further, very soft springs (shown in Fig. 12.3 by dashed lines) are entered into the diagram. These springs connect the fictitious nodes to the utmost nodes of lateral branches. Thus, the ramified system is transformed into a tape system. Sequential numbering is assigned to its nodes including the fictitious ones. The node-disk number is denoted by s and the total number of nodes by S

$$s = 1, 2, \ldots, S-1, S \tag{12.1}$$

where the designation $s-1, s$ is given to the spring connecting the $(s-1)$th and the s th disks.

Let us introduce some notation: j_s, $\vartheta_s(t)$ is the moment of inertia and rotation angle of the s th disk; $c_{s-1,s}$ is the torsional stiffness of the torsion bar connecting the $(s-1)$th and sth disks; g_s is the coefficient of viscous damping applied to the sth disk; $\mu_s(t)$ is the external periodic torque applied to the sth disk; $m_{s-1,s}(t)$ is the moment acting on the sth disk from the $(s-1, s)$th spring; and $0 \leq t \leq T$ is time.

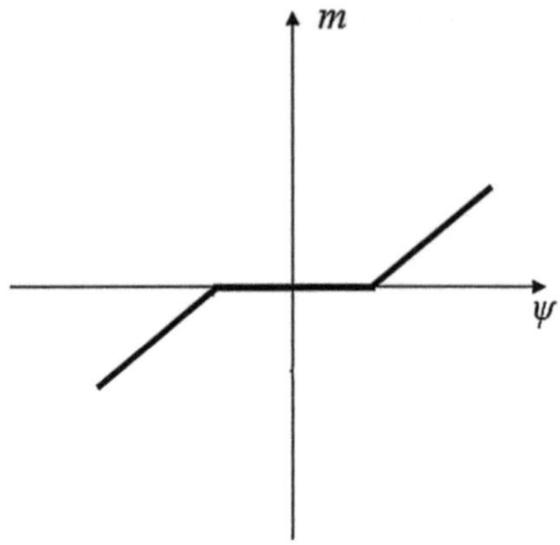

Fig. 12.4 The ideal function $m = c(\psi)$ in the presence of a backlash in the tooth connection of an ICE

The equation of oscillations of the sth disk is

$$-j_s \frac{\mathrm{d}^2}{\mathrm{d}t^2}\vartheta_s - g_s \frac{\mathrm{d}}{\mathrm{d}t}\vartheta_s + (m_{s,s+1} - m_{s-1,s}) + \mu_s = 0,$$
$$\vartheta_s(0) = \vartheta_s(T), \quad \frac{\mathrm{d}\vartheta_s}{\mathrm{d}t}(0) = \frac{\mathrm{d}\vartheta_s}{\mathrm{d}t}(T). \tag{12.2}$$

In this equation the linear relationship $m_{s-1,s} = c_{s-1,s}\psi_{s-1,s}$ holds for the torsion bar, where

$$\psi_{s-1,s} = \vartheta_s - \vartheta_{s-1}. \tag{12.3}$$

The function $m = c(\psi)$ is nonlinear regarding the connection of gearwheels. An ideal graph of this connection is shown in Fig. 12.4. The connection is bilateral because a tooth of one contact gearwheel is located between two teeth of the other adjacent gearwheel.

Let us introduce the vector functions $\vartheta = (\vartheta_s)$, $\mu = (\mu_s)$, which allows us to write concisely

$$L(\vartheta) = A(\vartheta) + \mu = 0. \tag{12.4}$$

instead of the system of Eq. (12.2).

This operator equation can be solved using the Newton–Kantorovich method. The iterative algorithm can be written as

$$A'(\vartheta^n)(\vartheta^{n+1} - \vartheta^n) + L(\vartheta^n) = 0, \tag{12.5}$$

where

$$A' = A'(\vartheta) = \frac{\mathrm{d}A(\vartheta)}{\mathrm{d}\vartheta}; \tag{12.6}$$

and n is the number of approximation.

Otherwise

$$A'\vartheta^{n+1} = q(\vartheta^n), \tag{12.7}$$

where

$$q(\vartheta^n) = A'\vartheta^n - L(\vartheta^n) \tag{12.8}$$

At the $(n+1)$th step of the iterative process (12.7) we consider the functions ϑ^n and $q(\vartheta^n)$ to be known. There is a need to determine ϑ^{n+1} (the desired function in the $(n+1)$th approximation) by the recurrent formula (12.7). To do this the index n can be omitted since the algorithm for solution of Eq. (12.7) does not depend on this index

$$A'\vartheta = q \tag{12.9}$$

or, in more detail, taking Eq. (12.2) into account

$$-j_s\frac{\mathrm{d}^2}{\mathrm{d}t^2}\vartheta_s - g_s\frac{\mathrm{d}}{\mathrm{d}t}\vartheta_s + c_{s,s+1}(\vartheta_{s+1} - \vartheta_s) - c_{s-1,s}(\vartheta_s - \vartheta_{s-1}) - q_s = 0. \tag{12.10}$$

If the $(s-1)$th and sth disks are connected by a torsion bar, then $c_{s-1,s}$ represents the torsional stiffness. Linearized stiffness for the connection of gearwheels is

$$c_{s-1,s} = c_{s-1,s}(\psi_{s-1,s}) = \frac{dm_{s-1,s}(\psi_{s-1,s})}{\mathrm{d}\psi_{s-1,s}}, \psi_{s-1,s} = \vartheta_s - \vartheta_{s-1}. \tag{12.11}$$

Approximate calculation of this derivative can be done by smoothing the function depicted in Fig. 12.4 and then replacing it with the function given in Fig. 12.5. This should be done because the Newton–Kantorovich method does not allow the use of functions whose first derivative is zero and whose second derivative is not bounded, as is the case for the ideal function shown in Fig. 12.4. Smoothing can partly be justified by the fact that there is always a layer of oil between the teeth. A graph of function $c = c(\psi)$ is plotted in Fig. 12.5.

Equation (12.10) can be solved using the projection method again. The projection conditions can be written as

$$\frac{1}{T}\int_0^T \mathrm{e}^{-ik\lambda t}\left(-j_s\frac{\mathrm{d}^2}{\mathrm{d}t^2}\vartheta_s - g_s\frac{\mathrm{d}}{\mathrm{d}t}\vartheta_s \right.$$

$$\left. + c_{s,s+1}(\vartheta_{s+1} - \vartheta_s) - c_{s-1,s}(\vartheta_s - \vartheta_{s-1}) - q_s\right)\mathrm{d}t = 0, \tag{12.12}$$

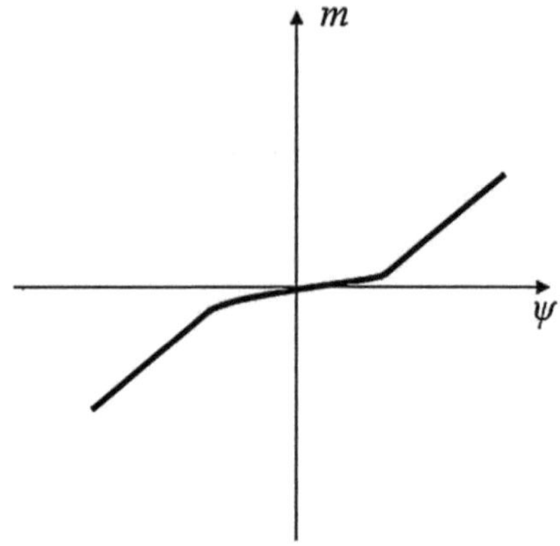

Fig. 12.5 The smoothed function $m = c(\psi)$ in the presence of a backlash in the tooth connection of an ICE

where

$$k = -K, -(K-1), \ldots, K-1, K, \quad \lambda = \frac{2\pi}{T}. \tag{12.13}$$

Calculating the integrals by parts and taking the periodicity of the harmonic functions and their derivatives into account, we can write that

$$\frac{1}{T}\int_0^T e^{-ik\lambda t}\left(-j_s \frac{d^2}{dt^2}\vartheta_s\right)dt = \frac{(k\lambda)^2 j_s}{T}\int_0^T e^{-ik\lambda t}\vartheta_s dt, \tag{12.14}$$

$$\frac{1}{T}\int_0^T e^{-ik\lambda t}\left(-g_s \frac{d}{dt}\vartheta_s\right)dt = \frac{-ik\lambda g_s}{T}\int_0^T e^{-ik\lambda t}\vartheta_s dt, \tag{12.15}$$

The desired solution can be approximately represented as a finite harmonic series

$$\vartheta_s = \sum_{m=-M}^{M} \vartheta_s^m e^{im\lambda t}, \quad M = K. \tag{12.16}$$

Substituting this series into the projection conditions (12.12) and formulas (12.14) and (12.15), followed by changing the order of integration and summation, we obtain a system of linear algebraic equations for the coefficients ϑ_s^k

$$\begin{aligned} &((k\lambda)^2 j_s - ik\lambda g_s)\vartheta_s^k \\ &+ \sum_{m=-M}^{M}\left(c_{s-1,s}^{km}\vartheta_{s-1}^m - (c_{s-1,s}^{km} + c_{s,s+1}^{km})\vartheta_s^m + c_{s,s+1}^{km}\vartheta_{s+1}^m\right) - q_s^k = 0, \end{aligned} \tag{12.17}$$

where, in particular, the equations

$$\frac{1}{T}\int_0^T e^{-ik\lambda t}c_{s-1,s}\vartheta_s dt = \frac{1}{T}\int_0^T e^{-ik\lambda t}c_{s-1,s}\sum_{m=-M}^{M}\vartheta_s^m e^{im\lambda t}dt = \sum_{m=-M}^{M}\vartheta_s^m c_{s-1,s}^{km}, \quad (12.18)$$

$$c_{s-1,s}^{km} = \frac{1}{T}\int_0^T c_{s-1,s}(t)e^{i(m-k)\lambda t}dt \quad (12.19)$$

are taken into account and the orthonormality property of the functions $e^{im\lambda t}$ (5.4) can also be used.

Entering the vectors $m_s = (q_s^k)$, $\theta_s = (\vartheta_s^k)$ and the Toeplitz matrices $C_{s-1,s} = (c_{s-1,s}^{km})$, we can write the system of Eq. (12.17) in vector-matrix form

$$A_{s,s-1}\theta_{s-1} + A_{ss}\theta_s + A_{s,s+1}\theta_{s+1} - m_s = 0, \quad (12.20)$$

where

$$A_{s,s-1} = C_{s-1,s}, \quad A_{s,s+1} = C_{s,s+1}$$

$$A_{ss} = \mathrm{diag}((k\lambda)^2 j_s - ik\lambda g_s) - (C_{s-1,s} + C_{s,s+1}).$$

In the band system considered the first and last nodes are not fixed against rotation; hence, the static component of the angular position of system nodes is uncertain. To avoid this, we adjoin "soft" springs (fastened at the other end) to the extreme nodes indicated, such that instead of (12.1) there will be $s = 0, 1, \ldots, , S, S+1$ and hence

$$\vartheta_0 = \vartheta_{S+1} = 0. \quad (12.21)$$

Solving the system of linear Eqs. (12.20) and (12.21) with a three-diagonal matrix can be made easier using the matrix sweep method which involves recursion formulas of the form (4.86). Recall that the band mechanical system considered was obtained by transforming the original branched system. This should be taken into account when drawing up the matrix sweep method algorithm. Necessary modification of the method of calculation can be carried out as follows. The nodes of the main chain, to which the lateral branches have been attached, are extracted. The force exerted by the side branch is added to these nodes during direct sweep. During reverse sweep the equality of displacements of the main chain node and attached fictitious node is taken into account.

Using this method we can calculate the vibrations of a real ICE (see Fig. 12.3 for a conditional scheme). The displacements of all units and forces in all connections for one oscillation period have been determined. For example, Figs. 12.6 and 12.7 show graphs of calculated moments acting on gear wheel 11 from the tooth connection of wheels 11 and 14 in both the absence and presence of backlashes.

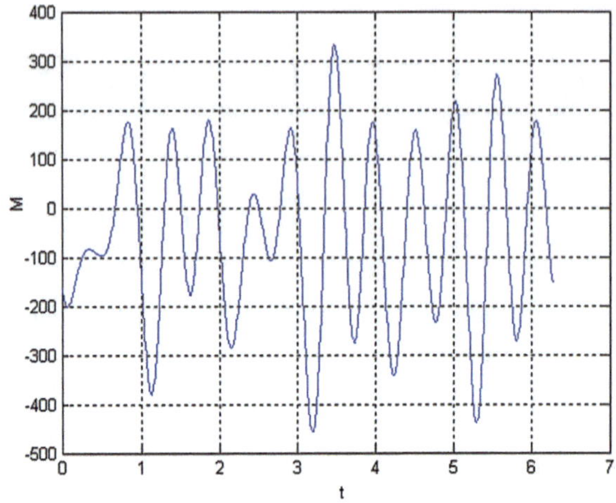

Fig. 12.6 Torque acting on node 11 from node 12 in the absence of backlashes in the tooth connections of an ICE gearbox (dimensions along the ordinate axis are given in meter dyne)

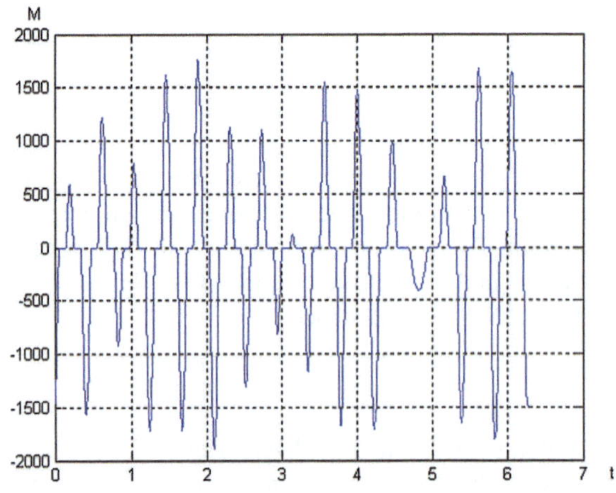

Fig. 12.7 Torque acting on node 11 from node 12 in the presence of backlashes in the tooth connections of an ICE gearbox (dimensions along the ordinate axis are given in meter dyne)

Comparison of the graphs shows that forces increase about 4.8 times as a result of backlashes in tooth connection. Calculation using a personal computer lasted just a few seconds.

References

1. Fridman, V., Tibbetts, D., & Piraner, I. (1997). Nonlinear chaotic and periodic vibrations in an internal combustion engine. In *Proceedings First International Conference on Control of Oscillations and Chaos* (Vol. 3, pp. 403–406).

Appendix A
The Operator Notation of Equations of the Theory of Elasticity and Boundary Conditions in Curvilinear Coordinates

The equilibrium equations and elasticity relations at static loading and harmonic vibrations of a three-dimensional body, shell and rod in curvilinear coordinates can be written as

$$D\xi + A\eta - f = 0, \quad D^{*}\eta - B\xi - e = 0, \tag{A.1}$$

where

$$A = C - \lambda^{2}R; \quad \text{and} \quad \lambda \, \text{is oscillation frequency.} \tag{A.2}$$

(1) In the case of a three-dimensional body [1, 2] in these equations $\xi = \tau^{ij}$ is the stress tensor; $\eta = u_i$ is the displacement vector; $f = f^i$ is the vector of volumetric forces; and $e = e_{ij}$ is the strain tensor that does not satisfy the conditions of continuity (dislocation strain tensor).

The components of tensors and vectors can be attributed to the oblique-angled coordinate system α_i ($i = 1, 2, 3$), in which the superscript denotes contravariance and the subscript covariance. Further, D and D^{*} are differential operators defined by the formulas

$$D\xi = -\nabla_i \tau^{ij}, \quad D^{*}\eta = \frac{1}{2}(\nabla_i u_j + \nabla_j u_i), \tag{A.3}$$

where ∇_i is the operation of covariant differentiation.

In Eq. (A.1) A and B are algebraic operators such that

V. Fridman, *Theory of Elastic Oscillations*, Foundations of Engineering Mechanics, DOI 10.1007/978-981-10-4786-2

$$B\xi = b_{ijkl}\tau^{kl}, \tag{A.4}$$

where b_{ijkl} is the tensor of elastic constants. For an isotropic body

$$b_{ijkl} = -\frac{\nu}{E} g_{ij} g_{kl} + \frac{1+\nu}{2E}(g_{ik}g_{jl} + g_{jk}g_{il}), \tag{A.5}$$

where E is the modulus of elasticity; ν is the ratio of transverse compression; and g_{ij} is the metric tensor.

In formula (A.2) $C = c^{ij}$ is the tensor of coefficients of elasticity of the medium that contains the deformable body

$$R = \rho g^{ij}, \quad \rho \text{ is specific mass.} \tag{A.6}$$

(2) The theory of shells, based on Kirchhoff's hypotheses [3], gives us in Eq. (A.1)

$$\xi = (S, H), \quad \eta = (u, w), \quad f = \begin{pmatrix} t \\ p \end{pmatrix}, \quad e = \begin{pmatrix} g \\ k \end{pmatrix},$$

where $S = S^{\mu\nu}, H = H^{\mu\nu}$ are the symmetric tensors of forces and moments in the system of curvilinear coordinates on the middle surface of the shell $\alpha_\mu (\mu = 1, 2)$; $u = u_\mu$ is the displacement vector in the tangent plane; w is normal displacement; $t = t^\mu$ is the external load vector in the tangent plane; p is normal load; $g = g_{\mu\nu}$ is the dislocation strain tensor of tension and shearing of the middle surface; and $k = k_{\mu\nu}$ is the tensor of dislocation change in the curvature of the middle surface

$$D\xi = \begin{pmatrix} -\nabla_\nu S^{\nu\mu} + & 2b^\mu_\nu \nabla_\kappa H^{\nu\kappa} + H^{\nu\kappa}\nabla_k b^\mu_\nu \\ -b_{\mu\nu}S^{\mu\nu} + & b^\nu_\mu b_{\nu\mu} H^{\mu\kappa} - \nabla_\mu \nabla_\nu H^{\mu\nu} \end{pmatrix}, \tag{A.7}$$

$$D^*\eta = \begin{pmatrix} \theta_{11} & \theta_{12} \\ \theta_{21} & \theta_{22} \end{pmatrix},$$

$$\theta_{11} = \frac{1}{2}(\nabla_\mu u_\nu + \nabla_\nu u_\mu), \quad \theta_{12} = -b_{\mu\nu} w, \tag{A.8}$$

$$\theta_{21} = -\frac{1}{2}\left(b^\kappa_\mu \nabla_\nu u_\kappa + b^\kappa_\nu \nabla_\mu u_\kappa + \nabla_\mu (b^\kappa_\nu u_\kappa) + \nabla_\nu (b^\kappa_\mu u_\kappa)\right),$$
$$\theta_{22} = (b^\kappa_\mu b^\kappa_{\kappa\nu} + b^\kappa_\nu b^\kappa_{\mu\kappa}) w + \nabla_\mu \frac{\partial w}{\partial \alpha^\nu} - \nabla_\nu \frac{\partial w}{\partial \alpha^\mu}.$$

For the algebraic operators B, C and R we have

$$B\xi = \begin{pmatrix} h^{-1} b_{\kappa\delta\mu\nu} S^{\kappa\delta} & 0 \\ 0 & 12h^{-3} b_{\kappa\delta\mu\nu} H^{\kappa\delta} \end{pmatrix}, \tag{A.9}$$

$$C\eta = \begin{pmatrix} c^{\mu\nu} u_\nu \\ cw \end{pmatrix}, \quad R = \rho a^{\mu\nu} h,$$

where h is the thickness of the shell;

$$b_{\kappa\delta\mu\nu} = -\frac{\nu}{E} a_{\kappa\delta} a_{\mu\nu} + \frac{1+\nu}{2E} (a_{\kappa\mu} a_{\delta\nu} + a_{\delta\mu} a_{\kappa\nu}); \tag{A.10}$$

$a_{\mu\nu}$ is the metric tensor of the middle surface; ∇_μ is the operation of differentiation on the surface; $b_{\mu\nu}$ are coefficients of the second quadratic form of the middle surface of the shell; $c^{\mu\nu}$ is the tensor of elastic coefficients of foundation at displacement in the tangent plane; c is the coefficient of elasticity of foundation at normal displacement; and ρ is the specific mass of the shell.

(3) The equilibrium equations and elasticity relations for a curvilinear rod can be conveniently recorded in the moving rectangular coordinate system $\zeta_i (i = 1, 2, 3)$, where ζ_1 is the axis tangential to the rod axis, and the axes ζ_2 and ζ_3 coincide with the directions of the principal axes of inertia of the rod [4].

The position of the trihedral vertex can be defined by coordinate α, counting along the axis of the rod. The trihedral angular velocity of rotation at motion of its vertex along the axis of the rod with unit velocity can be denoted by $\omega = \omega_i$.

Equations (A.1) and (A.2) give us

$$\xi = (Q, M), \eta = (u, \vartheta), \quad f = \begin{pmatrix} p \\ m \end{pmatrix}, \quad e = \begin{pmatrix} g \\ k \end{pmatrix},$$

where $Q = Q^i$ is force; $M = M^i$ is cross-sectional moment in the rod; $u = u_i$ is displacement of the point of the rod; $\vartheta = \vartheta_i$ is turning of the cross-sectional plane of the rod under deformation; $p = p^i$ isexternal distributed load on the rod; $m = m^i$ isexternal distributed moment; $g = g_i$ is the dislocation strain of tension and shearing; and $k = k_i$ is the dislocation strain of torsion and bending of the rod.

For curvilinear rods

$$D\xi = \begin{pmatrix} -\frac{dQ^i}{d\alpha} - \varepsilon_k^{ij} \omega_j Q^k \\ \varepsilon_k^{ij} \tau_j Q^k - \frac{dM^i}{d\alpha} - \varepsilon_k^{ij} \omega_j M^k \end{pmatrix}, \tag{A.11}$$

where τ_i is the unit vector of the tangent to the rod axis; and ε_k^{ij} is an absolutely antisymmetric tensor [5].

Further,

$$D^*\eta = \begin{pmatrix} \frac{du_i}{d\alpha} + \varepsilon_i^{kj}\omega_j u_k + \varepsilon_i^{kj}\tau_j\vartheta_k \\ \frac{d\vartheta_i}{d\alpha} + \varepsilon_i^{kj}\omega_j\vartheta_k \end{pmatrix}, \tag{A.12}$$

$$B\xi = \begin{pmatrix} b_{ij}Q^j \\ \beta_{ij}M^j \end{pmatrix}, \quad C\eta = \begin{pmatrix} c^{ij}u_j \\ \gamma^{ij}\vartheta_j \end{pmatrix}, \quad R\eta = \begin{pmatrix} \rho^{ij}u_j \\ J^{ij}\vartheta_j \end{pmatrix}, \tag{A.13}$$

where b_{ii} are the compliances of a rod at tension ($i = 1$) and at shearing ($i = 2, 3$), $b_{ij} = 0$ ($i \neq j$); β_{ii} are the compliances of a rod at torsion ($i = 1$) and at bending ($i = 2, 3$), $\beta_{ij} = 0$ ($i \neq j$); c^{ii} are the coefficients of elasticity of foundation under longitudinal ($i = 1$) and transversal ($i = 2, 3$) displacement of the point of the axis of the rod, $c^{ij} = 0$ ($i \neq j$); γ^{ii} are the coefficients of elasticity of foundation at turning around the rod axis ($i = 1$) and around the cross-sectional inertia axes ($i = 2, 3$), $\gamma^{ij} = 0$ ($i \neq j$); J^{ii} are moments of inertia per unit length of the rod relative to the axis of the rod ($i = 1$) and to the axes of inertia of the cross-section ($i = 2, 3$), $J^{ij} = 0$ ($i \neq j$), $\rho^{ij} = 0$ ($i \neq j$), $\rho^{ii} = \rho$; and ρ is mass per unit length of the rod.

The specified and unknown force factors, displacements and operators entered in Eq. (A.1) are functions of the point of the domain Ω (occupied by a three-dimensional elastic body, a shell middle surface and a rod axis).

Conditions at the boundary Γ (on the surface of a three-dimensional body, at the edge of a shell, at the ends of the rod) should be added to Eq. (A.1)

$$\begin{matrix} X - F = 0 \quad \text{on} \quad \Gamma_1(X = N\xi), \\ Y - E = 0 \quad \text{on} \quad \Gamma_2(Y = T\eta), \end{matrix} \quad (\Gamma_1 + \Gamma_2 = \Gamma). \tag{A.14}$$

(1) In the three-dimensional problem X is the stress vector at the site of the body surface with the unit vector of the normal $n = n_j$; Y is the displacement vector of points on the body surface; and F and E are given values for these vectors.

Relations (A.14) in the three-dimensional problem are of the form

$$X^k = t_i^k n_j \tau^{ij}, \quad Y_k = t_k^i u_i, \tag{A.15}$$

where t_i^k, t_k^i are tensors of the components of the basis vectors; and X^k, Y_k are vectors of the stress and displacement relating to the rectangular coordinate system that is associated with the surface point.

(2) In the theory of shells

$$X = (Q, L), \quad Y = (U, \theta), \tag{A.16}$$

where $Q = Q^i$ is the vector of force at the edge of the shell; L is the bending moment at the edge of the shell; $U = U_i$ is the displacement vector at the edge of the shell; and θ is the turning angle of the shell edge.

The projections on the axes of the natural trihedron associated with the edge are

$$\begin{aligned} Q^1 &= \nu_\mu \nu_\nu (S^{\mu\nu} - b^\mu_\kappa H^{\mu\kappa}) - k\nu_\mu t_\nu H^{\mu\nu}, \quad k = t^\mu \nu_\nu H^{\mu\nu}, \\ Q^2 &= \nu_\mu t_\nu (S^{\mu\nu} - b^\nu_\kappa H^{\mu\kappa}) - l\nu_\mu t_\nu H^{\mu\nu}, \quad l = t^\mu t_\nu b^\nu_\mu, \\ Q^3 &= \nu_\mu \nabla_\nu H^{\mu\nu} + \frac{d}{ds}(\nu_\mu t_\nu H^{\mu\nu}), \quad L = \nu_\mu \nu_\nu H^{\mu\nu} \end{aligned} \tag{A.17}$$

where the natural trihedron is formed with the tangential normal, tangent and normal to the middle surface; ν_μ, t_μ are unit vectors of the tangential normal and the tangent; and s is a coordinate counted along the boundary contour. Further

$$U_1 = u_1, \quad U_2 = u_2, \quad U_3 = u_3, \quad \theta = -\nu^\kappa \left(\frac{\partial w}{\partial \alpha^\kappa} + b^\nu_\kappa u_\nu \right). \tag{A.18}$$

Relations (A.16)–(A.18) impart a concrete form to the relationships (A.14).

(3) In the case of a curvilinear rod

$$X = (Q, M), \quad Y = (U, \theta), \tag{A.19}$$

where $Q = Q^i$ is force; $M = M^i$ is moment at the ends of rod; $U = u_i$ is displacement; $\theta = \vartheta_i$ is turning of the rod ends; and F and E are prescribed values of X and Y.

We conclude our consideration of boundary conditions by noting that further reasoning could also be extended to the case when some force factor projections and all generalized displacement projections are given on the part of the boundary Γ. This would require partitioning the border into parts Γ_1 and Γ_2 for each projection separately by just computing integrals over the boundary Γ.

Moreover, for completeness we consider boundary conditions that are more general in nature than (A.14); namely

$$X + KY - F = 0 \quad \text{on} \quad \Gamma_1, \quad Y + PX - E = 0 \quad \text{on} \quad \Gamma_2. \tag{A.20}$$

These conditions apply when a three-dimensional body is surrounded by an elastic layer, the edge of a shell and the ends of a rod (which are elastically supported and elastically clamped). In relations (A.20) the linear bounded operators K and P characterize, respectively, the stiffness and compliance of the elastic layer, elastic support and sealing.

Let us highlight some properties of the operators entering into the system of equations (A.1) and boundary conditions (A.14) and (A.20). To do this, we introduce scalar products:

(1) in the three-dimensional problem

$$\int\limits_{\Omega} \xi e \mathrm{d}\Omega = \int\limits_{\Omega} \tau^{ij} e_{ij} \mathrm{d}\Omega, \int\limits_{\Omega} f\eta \mathrm{d}\Omega = \int\limits_{\Omega} f^i u_i \mathrm{d}\Omega,$$
$$\int\limits_{\Gamma} XY \mathrm{d}\Gamma = \int\limits_{\Gamma} X^i Y_i \mathrm{d}\Gamma, \tag{A.21}$$

(2) in the shell deformation problem

$$\int\limits_{\Omega} \xi e \mathrm{d}\Omega = \int\limits_{\Omega} (S^{\mu\nu}\varepsilon_{\mu\nu} + H^{\mu\nu}\beta_{\mu\nu}) \mathrm{d}\Omega,$$
$$\int\limits_{\Omega} f\eta \mathrm{d}\Omega = \int\limits_{\Omega} (t^{\mu} u_{\mu} + pw) \mathrm{d}\Omega, \int\limits_{\Gamma} XY \mathrm{d}\Gamma = \int\limits_{\Gamma} (Q^i U_i + L\theta) \mathrm{d}\Gamma, \tag{A.22}$$

(3) in the curvilinear rod deformation problem

$$\int\limits_{\Omega} \xi e \mathrm{d}\Omega = \int\limits_{\Omega} (Q^i g_i + M^i k_i) \mathrm{d}\Omega,$$
$$\int\limits_{\Omega} f\eta \mathrm{d}\Omega = \int\limits_{\Omega} (p^i u_i + m^i \vartheta_i) \mathrm{d}\Omega, \int\limits_{\Gamma} XY \mathrm{d}\Gamma = \int\limits_{\Gamma} (Q^i u_i + M^i \vartheta_i) \mathrm{d}\Gamma. \tag{A.23}$$

It can be shown that the operators D and D^* are conjugate in the sense of Lagrange in every case considered [6]

$$\int\limits_{\Omega} D\xi\eta \mathrm{d}\Omega = \int\limits_{\Omega} \xi D^* \eta \mathrm{d}\Omega - \int\limits_{\Gamma} XY \mathrm{d}\Gamma. \tag{A.24}$$

The operators B and R are positive, while the operators C, K and P are nonnegative

$$\int_{\Omega} R\eta\eta \mathrm{d}\Omega > 0(\eta \neq 0), \int_{\Omega} \xi B\xi \mathrm{d}\Omega > 0(\xi \neq 0), \int_{\Omega} C\eta\eta \mathrm{d}\Omega \geq 0,$$
$$\int_{\Gamma_1} KYY \mathrm{d}\Gamma \geq 0, \int_{\Gamma_2} XPX \mathrm{d}\Gamma \geq 0. \tag{A.25}$$

Other scalar products can also be used

$$\int_{\Omega} ee^* \mathrm{d}\Omega = \int_{\Omega} e_{ij} e^{ij} \mathrm{d}\Omega, \quad \int_{\Omega} YY^* \mathrm{d}\Omega = \int_{\Omega} Y_i Y^i \mathrm{d}\Omega,$$

$$\int_{\Omega} ff^* \mathrm{d}\Omega = \int_{\Omega} f^i f_i \mathrm{d}\Omega, \quad \int_{\Omega} XX^* \mathrm{d}\Omega = \int_{\Omega} X^i X_i \mathrm{d}\Omega$$

and so on.

References

1. Lurie, A. I. (2005). *Theory of elasticity* (1050 p). Berlin: Springer.
2. Sneddon, J. N., & Berry, D. S. (1958). *The classical theory of elasticity* (219 p). Berlin–Gottingen–Heldelberd: Springer.
3. Naghdi, P. M. (1963). Foundations of elastic shell theory. *Progress in solid mechanics* (Vol. 4, 193 p). Amsterdam, North–Holland Publishing Company.
4. Lurie, A. I. (1941). On small deformations of curvilinear rods. In *Proceedings of the Leningrad Polytechnic Institute* (No 3, pp. 38–50) (in Russian).
5. Mc. Connell A. J. (1957). *Application of tensor analysis* (411 p). N.Y.: Dover Publications, Inc.
6. Fridman, V. M., & Chernina, V. S. (1969). *The modified Galerkin-Ritz method, Associated with the mixed variational principle in the theory of elasticity* (No 1, pp. 64–78). The Academy of Sciences of the USSR, Mechanics of Solids.

V. Fridman, *Theory of Elastic Oscillations*, Foundations of Engineering Mechanics, DOI 10.1007/978-981-10-4786-2

MIX
Papier aus verantwortungsvollen Quellen
Paper from responsible sources
FSC® C105338
FSC
www.fsc.org

If you have any concerns about our products,
you can contact us on
ProductSafety@springernature.com

In case Publisher is established outside the EU,
the EU authorized representative is:
Springer Nature Customer Service Center GmbH
Europaplatz 3, 69115 Heidelberg, Germany

Printed by Libri Plureos GmbH
in Hamburg, Germany